The ASQ
Six Sigma Black Belt
Pocket Guide

Also available from ASQ Quality Press:

The Certified Six Sigma Master Black Belt Handbook
T.M. Kubiak

The Certified Six Sigma Black Belt Handbook, Second Edition
T.M. Kubiak and Donald W. Benbow

The Certified Six Sigma Green Belt Handbook
Roderick A. Munro, Matthew J. Maio, Mohamed B. Nawaz,
Govindarajan Ramu, and Daniel J. Zrymiak

*Six Sigma Green Belt, Round 2: Making Your Next Project
Better than the Last One*
Tracy L. Owens

*Lean Acres: A Tale of Strategic Innovation and Improvement
in a Farm-iliar Setting*
Jim Bowie

Six Sigma for the New Millennium: A CSSBB Guidebook,
Second Edition
Kim H. Pries

The Quality Toolbox, Second Edition
Nancy R. Tague

Mapping Work Processes, Second Edition
Bjørn Andersen, Tom Fagerhaug, Bjørnar Henriksen, and
Lars E. Onsøyen

Root Cause Analysis: Simplified Tools and Techniques,
Second Edition
Bjørn Andersen and Tom Fagerhaug

*Root Cause Analysis: The Core of Problem Solving and
Corrective Action*
Duke Okes

To request a complimentary catalog of ASQ Quality Press
publications, call 800-248-1946, or visit our website at http://
www.asq.org/quality-press.

The ASQ
Six Sigma Black Belt
Pocket Guide

T. M. Kubiak

ASQ Quality Press
Milwaukee, Wisconsin

American Society for Quality, Quality Press, Milwaukee 53203
© 2014 by ASQ
All rights reserved. Published 2013
Printed in the United States of America
21 20 19 18   8 7 6 5 4

Library of Congress Cataloging-in-Publication Data

Kubiak, T. M.
    The ASQ six sigma black belt pocket guide / T. M. Kubiak.
        pages cm
    Includes bibliographical references and index.
    ISBN 978-0-87389-856-0 (soft cover : pocket guide)

    ISBN 978-1-63694-138-7 (paperback)

    1. Quality control—Statistical methods—Handbooks, manuals, etc. I. Title.

    TS156.K8235 2013
    519.8'6—dc23   2013041451

ASQ Mission: The American Society for Quality advances individual, organizational, and community excellence worldwide through learning, quality improvement, and knowledge exchange.

Attention Bookstores, Wholesalers, Schools, and Corporations: ASQ Quality Press books, video, audio, and software are available at quantity discounts with bulk purchases for business, educational, or instructional use. For information, please contact ASQ Quality Press at 800-248-1946, or write to ASQ Quality Press, P.O. Box 3005, Milwaukee, WI 53201-3005.

To place orders or to request ASQ membership information, call 800-248-1946. Visit our website at http://www.asq.org/quality-press.

Portions of the input and output contained in this publication/book are printed with permission of Minitab Inc. All material remains the exclusive property and copyright of Minitab Inc. All rights reserved.

ASQ Quality Press
600 N. Plankinton Ave.
Milwaukee, WI 53203-2914
Email: books@asq.org
Excellence Through Quality™

For Nico, Andriani, Maria Christina, and Elias Tsioutsias—family through the good, through the bad, and forever.

Table of Contents

List of Figures and Tables

Preface

I am pleased to provide our readers with the first edition of *The ASQ Six Sigma Black Belt Pocket Guide*.

As you read this book, you may wonder why I have chosen to use "Lean Six Sigma" in place of "Six Sigma" in most instances. The answer is simple: both the Black Belt and Master Black Belt bodies of knowledge discuss Lean, and the reality is that a practitioner must be competent in both. Further, the terminology also reflects the integrated nature of these tools and techniques.

This pocket guide assumes the reader has the necessary background and experience in quality and Lean Six Sigma, is already an ASQ Certified Six Sigma Black Belt, and has access to *The Certified Six Sigma Black Belt Handbook*, Second Edition. Ideally, the reader also has access to *The Certified Six Sigma Master Black Belt Handbook*. Consequently, it has been written at a high level to keep its size small and to retain its pocket guide status.

Unlike other pocket guides, this guide is designed specifically to address topics that I have found to cause problems, issues, and concerns for most Black Belts over the years. As such, its primary purpose is to serve as a useful reference guide for the Black Belt throughout his or her busy day, and particularly in meetings. See Chapter 1. It is not intended to be a tool guide like other pocket guides, or a preparation guide for the ASQ Black Belt examination.

That said, I have no doubt that it will nonetheless serve as a useful reference guide for both the ASQ Black Belt and Master Black Belt during certification examinations.

The Glossary for this pocket guide has intentionally been kept to a minimum and reflects mostly the terms used in the guide. A more detailed glossary is provided in either *The Certified Six Sigma Black Belt Handbook*, Second Edition or *The Certified Master Six Sigma Black Belt Handbook*.

A second Glossary has been included. This short glossary is limited to the most common Japanese terms used by quality and Lean Six Sigma professionals.

Additional glossary elements including statistical tables have been included so that they are readily at your fingertips. Unfortunately, some tables had to be excluded due to their size. Fortunately, you can find them in either of the above handbooks.

Suggestions for improving this pocket guide may be sent to authors@asq.org.

I hope you find this pocket guide a useful aid in your daily work.

—T. M. Kubiak

Acknowledgments

This was a particularly difficult project, and I could not have completed it successfully without much-needed support. The difficulty lay with determining criteria for selecting material to include such that this pocket guide is unique from all of the others on the market.

First, there was my wife, Darlene. Darlene served as my proofreader for the entire project. I am deeply appreciative of her patience and of her keen eyes. Of course, there was always a smile on her face as she eagerly pointed out my mistakes.

Second, special thanks go to Roderick Munro for his generous support, numerous suggestions, and encouragement while writing this pocket guide. His input brought clarity to me and helped me to determine how to craft the structure of this guide.

Third, I would like to express my deepest appreciation to Minitab Inc. for providing me with the Quality Companion 3 software and for permission to use several examples from this software, which was instrumental in creating several of the examples used throughout the book.

Last, I would like to thank the ASQ Quality Press management staff for their outstanding support and exceptional patience while I completed this project.

—T. M. Kubiak

Chapter 1

Introduction

This pocket guide assumes the reader has the necessary background and experience in quality and Lean Six Sigma, and has access to *The Certified Six Sigma Black Belt Handbook*, Second Edition. Ideally, the reader has access to *The Certified Six Sigma Master Black Belt Handbook* as well. Consequently, it has been written at a high level to keep its size small and to retain its pocket guide status.

Unlike other pocket guides, this guide is designed specifically to address topics that I have found to cause problems, issues, and concerns for most Black Belts over the years, based on my experience training them, as well as from many observations derived from user groups and online communities. As such, it is not intended to be a tool guide or a preparation guide for the ASQ Black Belt certification examination.

However, it is intended to serve as a useful reference tool for the Black Belt throughout his or her busy day. In addition, I expect it will be particularly useful in meetings by helping settle discussions and keeping the meetings moving.

As you read this book, you may wonder why I have chosen to use "Lean Six Sigma" in place of "Six Sigma" in most instances. The answer is simple: both the Black Belt and Master Black Belt bodies of knowledge discuss Lean, and the reality is that a practitioner must be competent in both. Further, the terminology also reflects the integrated nature of these tools and techniques.

CHAPTER CONTENT

Although there is an overall order to the chapter sequence and some relatively minor cross-referencing between chapters, this pocket guide has been designed so that each chapter is essentially independent of every other chapter. Therefore, you may feel comfortable starting at the beginning of any chapter you desire. Below is a brief synopsis of each.

Chapter 2 provides a detailed explanation of the DMAIC methodology and defines the purpose of each phase. A detailed flowchart of the DMAIC methodology is provided that includes the tollgates and illustrates how a failure at each tollgate can send the project backward one or more phases.

Chapter 3 identifies the various tools and techniques commonly used in *Design for Six Sigma* (DFSS). Also included is a discussion of various DFSS methodologies and a side-by-side comparison of each by phase. This chapter will help the Black Belt understand the relative strengths and weaknesses of each of four DFSS methodologies.

Chapter 4 is an essential chapter that assists the Black Belt in distinguishing among project methodologies. More specifically, a flowchart is provided that allows the Black Belt to determine which methodology to follow, such as project management, DFSS, Lean, Six Sigma.

Chapter 5 details the elements of the project charter, and examples are provided as appropriate. Significant attention has been given to scoping a project.

Chapter 6 clarifies the concepts of project savings. Both hard and soft savings are defined. Detailed flowcharts and a worksheet are provided to help the Black Belt in determining project savings. This chapter will be particularly useful during those highly argumentative project meetings.

Chapter 7 is a relatively simple chapter that identifies commonly used tools and techniques by each phase of DMAIC. While thorough, it is far from complete.

Chapter 8 identifies the Lean Six Sigma roles and responsibilities commonly used in a deployment. While there is no set agreement among all Lean Six Sigma practitioners, these roles and responsibilities are considered basic.

Chapter 9 addresses stakeholder engagement and communication. This topic is a paramount issue in all deployments and in every project. Examples of stakeholder analysis and communication forms are provided.

Chapter 10 discusses tollgates and the tollgate review process. Tollgates are often short-changed by champions. This chapter should help you help your champion get it right. If done correctly, tollgates add value.

Chapter 11 details the role of the Lean Six Sigma coach and differentiates the role of the coach from that of the mentor. Further, cost associated with the role of the coach is addressed.

Chapter 12 addresses the use of the process map and how to construct one properly. Unlike other chapters in this guide, this chapter is more of a teaching chapter. It is intended to demonstrate how to get the most out of a process map.

Chapter 13 is a significant chapter for the Black Belt who must deal with data collection activities. This chapter provides the reader with information on how to develop operational definitions for data collection and metric development, as well as key considerations for developing customer-based data collection systems. A special section that delves into the details of data integrity and reliability has also been included. Key topics such as minimizing poor data accuracy, useful data collection techniques, and important data collection points have been addressed. In addition, seven data collection strategies have been suggested, along with the advantages and disadvantages of each.

Chapter 14 is a compilation chapter that addresses several different tools, including force-field analysis, accuracy versus precision, control charts, and team management. I specifically included team management because, in my experience, I have found that this is often a particularly weak area for many Black Belts. Black Belts like to deal with the technical aspects of a project but often fall short in managing the people side of the project. Failing to manage the team properly can frequently lead to the demise of a project.

The Appendix is filled with many relevant tables. However, I want to call your attention specifically to Appendices 1 and 2. These appendices depict the six-step process of project identification to project closure and are worth noting.

Several statistical tables have been included in the Appendix, with the exception of the F-table. Because of the need to include the F-table at many different alpha values, I specifically excluded this table due to the size impact it would have had on this pocket guide.

If you need an *F*-table, please use the ones in *The Certified Six Sigma Black Belt Handbook*, Second Edition or *The Certified Six Sigma Master Black Belt Handbook*, where tables have been created for both tails of the distribution. These tables allow the reader to avoid using the reciprocal relationship property that exists between the degrees of freedom and the chosen alpha value for the *F*-distribution.

Note: Appendix 11, the Glossary of Lean Six Sigma and Related Terms, is not as complete a glossary as you will find in the Black Belt and Master Black Belt handbooks. This particular glossary has been tailored to this pocket guide.

I hope you enjoy reading this pocket guide and find it useful in your daily work. Please feel free to forward suggestions to authors@asq.org.

Chapter 2

The DMAIC Methodology

INTRODUCTION

As all Lean Six Sigma professionals know, DMAIC (define, measure, analyze, improve, control) is the de facto methodology for improving processes using Lean Six Sigma. The reason for this is quite simple. The methodology is:

- Easy to understand

- Logical

- Complete

Further, it is sufficiently general to encompass virtually all improvement situations whether they are business or personal in nature.

DMAIC

Let's briefly explore the purpose at each phase of the methodology:

- *Define.* The purpose of this phase is to provide a compelling business case appropriately scoped, complete with SMART goals, and linked to a strategic plan.

- *Measure.* The purpose of this phase is to collect process performance data for primary and secondary metrics to gain insight and understanding into root causes and to establish performance baselines.

- *Analyze.* The purpose of this phase is to analyze and establish optimal performance settings for each X and verify root causes.

- *Improve.* The purpose of this phase is to identify and implement process improvement solutions.

- *Control.* The purpose of this phase is to establish and deploy a control plan to ensure that gains in performance are maintained.

Example 2.1

I have just received my annual performance appraisal, which was based on a 360-degree review process. I am not happy with the result since on a scale of 1–5 (5 being high), I averaged a 2.

However, rather than complain, I decide to take the result to heart and do something about it. After careful consideration, I plan to use the DMAIC approach as a means of driving personal improvement:

• *Define*. I am unhappy with my personal performance and want to move my average performance from a 2 to a 3 by the next annual performance appraisal.

• *Measure*. The 360-degree performance appraisal is very detailed and allows me to decompose my average result into various components that I can associate with particular behaviors. Each of these component parts has, in turn, an average result determined for them as well.

• *Analyze*. Since I am able to analyze the results of each of the component parts, I create a Pareto chart to determine which components have the greatest influence on my overall average score.

Consequently, I am able to determine the top five behaviors. This is an important consideration since I must balance my actual productive work time against my personal improvement time and work within the time allotted to my annual training plan.

• *Improve*. With the top five drivers of behavior identified, I set about to develop and execute a plan. In this case, my plan consists primarily of taking company-sponsored training and education courses.

• *Control*. As I take the courses, I diligently practice what I have learned and decide that it might be beneficial to keep a journal that details any particular interactions that might influence my performance ratings. This allows me to keep what I have learned in the foreground and to practice it continuously.

In addition, I decide to use intermediate 360-degree performance appraisals that "don't count" as I am the only one who will see the feedback. As long as the feedback on the top five that I am focusing on is improving, I am on track.

If not, I will need to revisit my plan and make appropriate changes. Furthermore, I am concerned that behaviors that were "in control" have not moved in the wrong direction. If any have, they will be included in the plan moving forward.

Although this is a very simplified example, it demonstrates how useful and easily applied DMAIC is, even in common or routine situations.

When Belts first learn about DMAIC, they, for the most part, see it depicted in a linear manner as shown in Figure 2.1. Unfortunately, this is an illusion that leads inexperienced belts to plow ahead from phase to phase, steamrolling weak champions/sponsors, particularly if the Belts

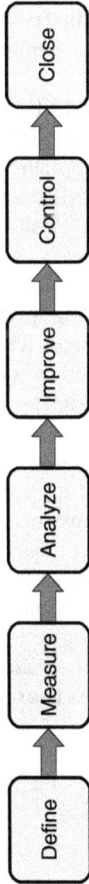

Figure 2.1 DMAIC in a dream world.

lack the support of a Master Black Belt coach. When this occurs, either projects fail or results are not sustained.

Figure 2.2 depicts DMAIC in the real world. As we all know, tollgate reviews occur after each phase. Diligent application of the proper tools and, above all, significant preparation are required to pass them. Passing them is not a given.

Notice that up to and including the *analyze* phase, the failure of a tollgate could set the project back one or two phases. This usually occurs when knowledge is gained in a phase contrary to expectations. In this case, the project team may be forced to retreat all the way back to *define* in order to restate the problem.

Beyond the analyze step, a failed tollgate usually results in correcting or performing additional work within the phase. However, at any phase, a project may be terminated. Generally, this will happen when projects are no longer synchronized with strategy, champions/sponsors have lost interest or been transferred, or the potential savings are less than expected.

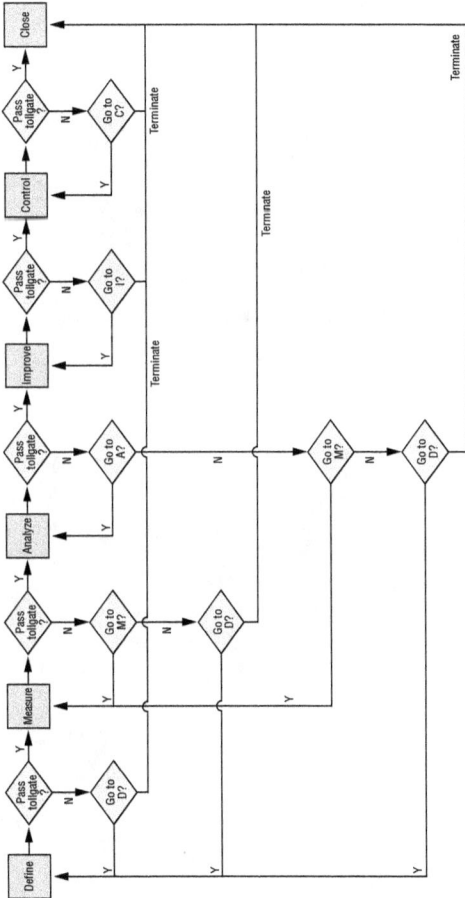

Figure 2.2 DMAIC in the real world.

Chapter 3

DFSS Methodologies

While DMAIC may be traditionally viewed as the foundation for Lean Six Sigma, its application is primarily limited to improving existing processes; it does little to address the design of new products or processes.

Fortunately, several additional structured methodologies exist. Several of the common methodologies are depicted in Figure 3.1. Each of these methodologies has its usefulness as long as the nuances of each are fully understood.

Table 3.1 has been created to provide a loose comparison between DMAIC and the DFSS methodologies. These comparisons are not exact, but do provide a sense of how the methodologies align. This table may be beneficial for readers who may be required to use multiple methodologies or find themselves in need of selecting one suitable for a particular project.

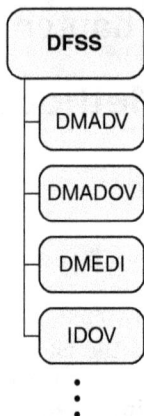

Figure 3.1   The DFSS family of methodologies.

Table 3.2 identifies some of the commonly used tools in DFSS, including the DFSS methodologies shown in Figure 3.1. However, this table also lists other tools that are useful regardless of the DFSS methodology chosen. Examples include axiomatic design and TRIZ.

DMADV

DMADV is a well-recognized Design for Six Sigma (DFSS) methodology and an acronym for

Table 3.1 Comparing DMAIC and DFSS methodologies.

DMAIC	DFSS			
	DMADV	**DMADOV**	**DMEDI**	**IDOV**
Define	Define	Define	Define	Identify
Measure	Measure	Measure	Measure	
Analyze	Analyze	Analyze	Explore*	Design
Improve	Design	Design	Develop	
		Optimize		Optimize
	Verify**	Verify**	Implement	Validate***
Control				

* Loosely aligned
** Often confused with "Validate"
*** Often confused with "Verify"

define, measure, analyze, design, and verify. Note that the ASQ Black Belt Body of Knowledge replaces "verify" with "validate." The difference is likely because, although different, "verify" and "validate" are often used synonymously.

The DMA portion of DMADV was addressed in Chapter 2, so we'll concentrate on the remaining DV portion:

• *Design*. Quite simply, this means carrying out the process of designing a product or process.

Table 3.2 Common tools used in DFSS.

Analytic hierarchical process (AHP)	Portfolio architecting
Axiomatic design	Process simulation
Critical parameter management	Pugh matrix
Design for X (DFX)	Quality function deployment (QFD)
DMADOV	Robust product design
DMADV	Statistical tolerancing
FMEA	Systematic design
IDOV	Tolerance design
Porter's five forces analysis	TRIZ

Many organizations have well-established policies and procedures for their respective design processes.

One valuable Lean Six Sigma technique that supports the design process is quality function deployment (QFD). Additional tools useful in this phase include pilot runs, simulations, prototypes, and models.

• *Verify.* This phase is directed at ensuring that the design output meets the design requirements and specifications, and is performed on the final product or process.

Verification means the design meets customer requirements and ensures that it yields the correct product or process. By contrast, *validation* speaks to the effectiveness of the design process itself and is intended to ensure that it is capable of meeting the requirements of the final product or process.

Both verification and validation are necessary functions in any design process. As such, this suggests that DMADV might be more appropriately named DMADVV.

DMADOV

The basic difference between DMADV and DMADOV is that "O" for "optimize" has been added. This may sound like a trivial observation, but many organizations' design processes do not include this refinement action. They are intended solely to produce a minimally workable product or process. DMADOV forces attention on the need to optimize the design.

Additional tools useful in this phase include design of experiments, response surface methodology (RSM), and evolutionary operations

(EVOP). These methods help the design team establish and refine design parameters.

DMEDI

The define–measure–explore–develop–imple-ment (DMEDI) methodology is appropriate where the limitations of DMAIC have been reached and customer requirements or expectations have not been met, or a quantum leap in performance is required.

Unlike DMAIC, DMEDI is not suitable for kaizen events, and it is generally more resource intensive, with projects taking considerably longer.

Let's explore each phase of DMEDI:

- *Define*. This phase is the same as in DMAIC.

- *Measure*. Unlike *measure* in DMAIC, this phase must emphasize the voice of the customer and the gathering of critical customer requirements since no baseline data are available. The quality function deployment (QFD) tool is valuable during this phase.

- *Explore*. This phase focuses on creating a high-level conceptual design of a new process that meets the critical customer requirements identified in the measure phase. Ideally, multiple designs are developed that provide options to be considered.

- *Develop.* The options developed in the explore phase are considered, and the "optimal" one is chosen based on its ability to meet customer requirements. Detailed designs are developed.

- *Implement.* During this phase, the new process may be simulated to verify its ability to meet customer requirements. In addition, a pilot is conducted, controls are established and put in place, and the new process is transferred back to the process owner.

IDOV

Unlike most of our other DFSS methodologies, IDOV contains only four phases and stands for identify–design–optimize–validate. Woodford ("Design for Six Sigma—IDOV Methodology") states that IDOV phases parallel the original MAIC phases.

However, a review of his descriptions does not confirm this. Table 3.1 has been modified to conform to the descriptions of each phase he provides.

Let's explore each phase of IDOV:

- *Identify.* This phase links the design to the voice of the customer via the customer, product, and technical requirements, establishes the business case, identifies CTQ variables, and conducts a competitive analysis. Quality function deployment will be a particularly useful tool during this phase.

- *Design.* This phase focuses on identifying functional requirements, deploying CTQs, developing multiple design concepts, and selecting the best-fit concept.

- *Optimize.* This phase focuses on developing a detailed design for the best-fit design chosen in the previous phase, predicting design performance, and optimizing the design. Additional work in this phase includes error-proofing, statistical tolerancing, prototype development, and optimization of cost.

- *Validate (verify).* During this phase, the prototype is tested and validated, and reliability, risks, and performance are assessed. Note: The acronym for IDOV is not consistently defined. The "V" is used about equally as often for "Validate" as it is for "Verify." Also, in many cases, when it is used as "Verify" in the acronym, it is explicitly described as "Validate" when the details of each letter are broken out.

Munro (2008) describes each phase of IDOV in more detail on pages 42–43 of *The Certified Six Sigma Green Belt Handbook*.

Chapter 4

Selecting the Proper Project Methodologies

Many times, the discussion of Six Sigma, DFSS, Lean, or even project management with senior management results in stares and a certain level of disinterest. Often, this is because they do not understand how these methodologies are separated or used. Consider this—how many times within your organization have all projects been considered Lean Six Sigma projects regardless of which methodology is appropriate?

Figure 4.1 illustrates how each methodology is chosen. Notice that project management has been included. It has been included, in particular, because many organizations have both project management and Lean Six Sigma organizations that seem to be in perpetual conflict with one another. Note: Figure 4.1 applies when we have to decide between the three methodologies (that is, project management, Lean, and Six Sigma), not when the answer to the problem is so

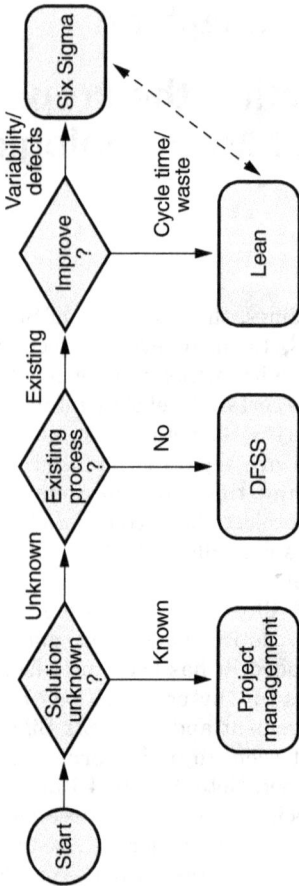

Figure 4.1 Selecting the proper project methodology.

straightforward and simple that we must simply just go do it.

Every Lean Six Sigma project leader at one time or another will find it necessary to employ the principles of project management. Each phase of the DMAIC methodology will likely see the start and close of many non–Lean Six Sigma mini projects as data are gathered, information is learned, or questions arise. For this reason alone, it is essential that the project leader understands these basic principles and uses them efficiently and effectively in the execution of the overall Lean Six Sigma project.

Remember, these mini projects should have known solution paths for employing the project management approach. Consequently, each step of the project can be defined and executed, resources can be defined, risks assessed, and a plan established. If this is done well, the desired results will follow.

Almost inevitably, the situation will arise during the course of a Lean Six Sigma project where one or more problems will surface that are adjacent or nearly adjacent to the process being worked on by the Lean Six Sigma project. The temptation will be great to adjust the scope of the project to include these problems. Resist it! Apply Figure 4.1 to the problems and spin off other project types. This approach permits better overall management of resources and projects.

Figure 4.1 also shows that any given Lean Six Sigma project may alternate between the use of Lean and Six Sigma. This is an important consideration as some organizations and practitioners tend to favor the use of one tool over another. Black Belts need to be well versed in the use of both sets of tools so that they can provide effective leadership and continuity to a project.

Chapter 5

Developing the
Project Charter

PROJECT CHARTER

The project charter, though a simple living
document, requires significant preparation time.
See Figure 5.1. It serves as a formal contract
between the champion and Black Belt, and helps
the team stay on track with the goals of the project and organization. There are typically six key
elements of a meaningful charter document:

- *Business case.* This identifies the dollars
  to be saved and establishes how the project
  aligns to the organization's strategies.
  It also states how the organization will
  fare better when the project reaches
  its goals. Although the business case
  generally appears first in a project charter,
  it is often easier to write the business
  case once the problem statement, goal
  statement, and project scope are defined.

Project identification related information	
Business case	Problem statement
Goal statement	Project scope
Project plan	Project team

Figure 5.1 A basic form of a charter document.

- *Problem statement.* This identifies what is working, what is not working, or where the pain point lies. If the pain point can not be identified, it is unlikely that a meaningful project can be defined. This critical statement is often glossed over.

- *Goal statement.* This identifies the project's objectives and targets and how the success of the project will be measured.

- *Project scope.* This specifies the boundaries of the project. The most common scope limitations are budget, time, authority, and resources. Additional aspects of the scope are addressed below.

- *Project plan.* The greater the amount of detail that can be provided, the better. However, an estimated completion date is required. Project durations must be determined so that, from a portfolio perspective, projects can be managed in terms of cost, schedule, and resources.

- *Project team.* The project team members should be identified along with their expected time contributions to the project. Project team members are knowledgeable and valuable resources who will likely be in demand for other projects as well.

Knowing the total demand requirements placed on team members is essential to ensure they are not overloaded and that their "regular" jobs do not suffer from their contributions to projects.

Furthermore, it is prudent to have charters replete with the appropriate signatures of authorization, not the least of which are the sponsor and financial representatives. The inclusion of the financial approval lends credibility and support to the proposed financial gains. This aspect is so important that many organizations establish financial approval as a requirement for a project to be considered for qualification.

The project sponsor has the ultimate accountability for developing the project charter. He or she will likely seek counsel from a Black Belt or Master Black Belt. However, in the end, senior leaders will look to the project sponsor for the success or failure of the project. It is for this reason that they need to be a major contributor to the project qualification process.

Figure 5.2 illustrates an alternate form of a project charter. Note how this form emphasizes baseline and goal performance and provides the opportunity to link the project directly to customer satisfaction.

Project Charter

Project Authorization

Organization:	Champion:	Process Owner:
Project: Project		Project #:
Problem Statement:		
Project Objective:		
Estimated Defect Level:	Initial Goal:	Estimated Benefits:
Approval Date:	Champion Signature:	Process Owner Signature:
Estimated Completion Date:	Project Leader:	Financial Analyst:

Project Team

Name	Role	Comments	Phone

Project Definition and Scoping

Metrics (unit of measure):

Critical to Satisfaction (linkage to customer):

Defect Definition (include opportunity):

Scope of Project:

Figure 5.2 Example of a project charter document.

Source: Courtesy of Minitab Quality Companion 3 software, Minitab Inc.

Goals and Benefits

Defect Levels/Goals:

	Date	DPMO(LT)	Zbench(ST)	Cpk
Baseline		0	0.00	0.00
Goal		0	0.00	0.00
Stretch Goal		0	0.00	0.00

Estimated Financial Benefits:

⚠ Important infomation

Hard Savings	$0
Soft Savings	$0
Implementation Costs	$0

Based on how many months: 12

Note: Improvement goals, estimated financial benefits, actual baseline DPMO, and Zbench should be reviewed and revised as needed after the end of the Measure phase when you have established a solid baseline for the project.

Measure phase completed on: _____

☐ Were goals revised after completion of Measure phase?

☐ Were financial benefits revised after completion of Measure phase?

Approved by Finance Representative:	Date of Finance Approval:

Figure 5.2 *Continued.*

PROBLEM STATEMENT

The problem statement should be a concise expla-nation of a current state that adversely impacts the enterprise. In many ways, the problem defi-nition phase is the most important phase of the DMAIC cycle. If this phase is not done thor-oughly, teams may move on to subsequent phases only to stall and cycle back through *define*. This phase should be emphasized, and teams should not move forward until the sponsor signs off on it.

Examples of well-defined problem statements:

- The reject rate for product X is so high that competitors are taking some of our market share.

- The cost of product B is so high it is making us noncompetitive.

Defining the problem statement is like finding the pain. If you can't find the pain in the statement, then the problem statement is not well written. In both of the examples above, the pain is well defined. In the first example, the pain is that competitors are taking some of the market share. In the second example, the high cost of product B is making the company lose its competitiveness.

GOAL STATEMENT

Goal statements must define at least one success (that is, primary) measure. However, baseline performance must be established for each primary measure. Frequently, primary measures are accompanied by secondary or countermeasures to ensure that when primary measures are improved, other measures (that is, secondary measures) do not suffer in the process. Likewise, baseline performance must be established for each secondary measure used.

Remember, goals should be SMART:

- *Specific*. This is not the place to be generic or philosophic. Nail down the goal.

- *Measurable*. Unless the team has measurable goals, they won't know whether they are making progress or whether they have succeeded.

- *Achievable*, yet aggressive. This is a judgment call, and experience with project planning and execution will help in meeting this requirement.

- *Relevant*. The goal must be specifically linked to the strategic goals of the enterprise.

- *Timely*. The goal must make sense in the time frame that the team has been given to work in.

An example of a SMART goal linked to the organization's strategic goals of customer satisfaction and quality might be:

- Reduce the reject rate for product X from its baseline of 45% to 5% within six months. Note: This is an absolute percentage reduction.

- Reduce the cycle time of product B from 90 days to less than 10 days 95% of the time within three months.

Remember, all processes can be measured in terms of quality (that is, defects) and cycle time. The cost of the process flows from these two metrics. Reducing cost without focusing on reducing defects and cycle time is a recipe for disaster.

SCOPING THE PROJECT

Lean Six Sigma projects sometimes suffer from a disagreement among the project team members regarding project boundaries. The process of defining scope, of course, can result in problems of extremes:

- Project definitions with scopes that are too broad may lead a team into a morass of connecting issues and associated problems beyond the team's resources. For example, "Improve customer satisfaction" with a complex product or service.

- Project boundaries that are set too narrow could restrict teams from finding root causes. For example, "Improve customer satisfaction by reducing variation in plating thickness" implies a restriction from looking at machining processes that may be the root cause of customer problems.

- The tendency is to err on the side of making the scope too broad rather than too narrow. Extra attention, effort, and time may be needed to ensure a proper scope. Don't shortcut the process.

- Several tools are available to assist in setting a project scope. These include:

  - *Pareto charts* to help in the prioritizing processes and sometimes in support of project justification

  - *Cause-and-effect diagrams* to broaden the thinking within set categories

  - *Affinity diagrams* to show linkages between the project and other projects or processes

  - *Process maps* to provide visualization and perhaps illuminate obvious project boundaries and relationships.

Note Table 7.1, which provides a more extensive list of tools commonly used in the define phase.

Collectively, these tools help "Belts" zero in on the scope and sometimes, even more importantly, what is out of scope.

Generally, there are two methods useful in defining projects' scopes. These are the:

- Dimensional method

- Functional method

Dimensional Method

This approach identifies an initial eight possible dimensions for characterizing a project scope:

- *Process.* Processes receive input from other processes and feed output to other processes. Tightly bound the process or subprocess associated with the project.

- *Demographics.* This would consider factors such as employee categorizations, gender, age, education, job level, and so on.

- *Relationships.* This would include such entities as suppliers, customers, contract personnel, and so on.

- *Organizational.* Which business units, divisions, sites, or departments are included?

- *Systems*. Which manual or computerized systems are included? For example, those using system A.

- *Geographical*. Which country or site is included?

- *Customer*. Which segmentation or category will be considered?

- *Combinations*. Any combination of the aforementioned.

Note: This list is subject to expansion. Readers are encouraged to submit ideas to authors@asq.org.

The advantage of this method is that it communicates the scope in straightforward and easy to understand terms. For example, the project scope will entail process A, employee category B, in site C, who produce products for customer D. Unfortunately, this method does not focus in on process inputs as well as the functional method.

In addition to defining what is in scope, it is often useful to define what is out of scope. Although one would seem to define the other, experience has shown that what is out of scope is frequently overlooked or not understood fully unless explicitly stated. Many sponsors, team members, and other stakeholders often fail to make this important connection. For example, process B is in scope, while processes A and C are out of scope.

Unwieldy scopes are one of the most frequent reasons cited for the demise of projects. When the scope is too large to be completed within the project plan time frame, or additional resources are not available, there may be a reluctance to go back and re-scope the project. When it is too small, projected savings may be overstated. Note: The project charter should be considered a living document and adjusted as information is gained and learning takes place.

Functional Method

The functional method is rooted more in logic and functional relationships between input and output variables than the dimensional method. The intent is to isolate critical input variables. Other names for this method include *process mapping decomposition*, x–y *diagrams*, Y = f(X), and *big* Y *exercises*.

This method is best illustrated by an example provided by Lynch, et al. (2003). The example deals with the issue of poor electric motor reliability. Figure 5.3 portrays the functional $Y = f(X)$ breakdown. Each of the X's should be supported by operational data and may be supported by subject matter expert opinions.

The breakdown continues, in this example, until the smallest meaningful part with continuous data is identified. Keep in mind that another Y from the brush reliability equation could have been identified as well. This would have yielded an additional, but concurrent, project.

The disadvantage of using this approach lies in the difficulty of explaining it to senior leaders and other levels of management.

Figure 5.4 is the same as Figure 5.3 but uses a process mapping decomposition breakdown format to isolate critical input variables. This format may be more appealing to senior leaders and management, in general.

Y_1—Electric motor reliability depends on:
X_1—***Motor reliability***
X_2—Controller reliability
X_3—Mechanical mounting integrity
X_4—Specific application or use

$$Y_1 = X_1 + X_2 + X_3 + X_4$$

Y_2—The reliability of the motor depends on the reliability of the following:
X_1—Stator reliability
X_2—Rotor reliability
X_3—***Brush reliability***
X_4—Housing reliability

$$Y_2 = X_1 + X_2 + X_3 + X_4$$

Y_3—Brush reliability depends on:
X_1—Assembly stack-up issues
X_2—Brush brittleness issues
X_3—Contamination of the brush assembly
X_4—Spring rate or condition issues
X_5—Brush dimensional issues
X_6—***Brush hardness issues***

$$Y_3 = X_1 + X_2 + X_3 + X_4 + X_5 + X_6$$

Y_4—Brush hardness issues depend on:
X_1—Mean brush hardness
X_2—***Variation in brush hardness***

$$Y_4 = X_1 + \boxed{X_2}$$

The method used in this example focused in on the variation in brush hardness. This now becomes the scope of the project.

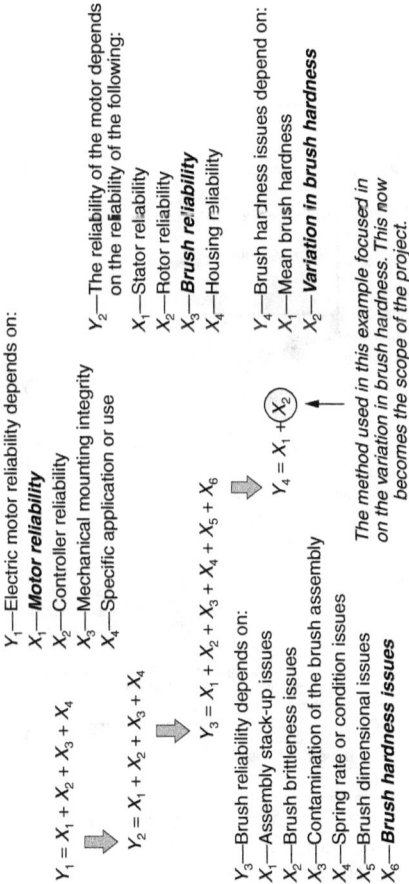

Figure 5.3   Example of project scoping using the functional method—$Y = f(X)$ format.

Source: Lynch et al. (2003).

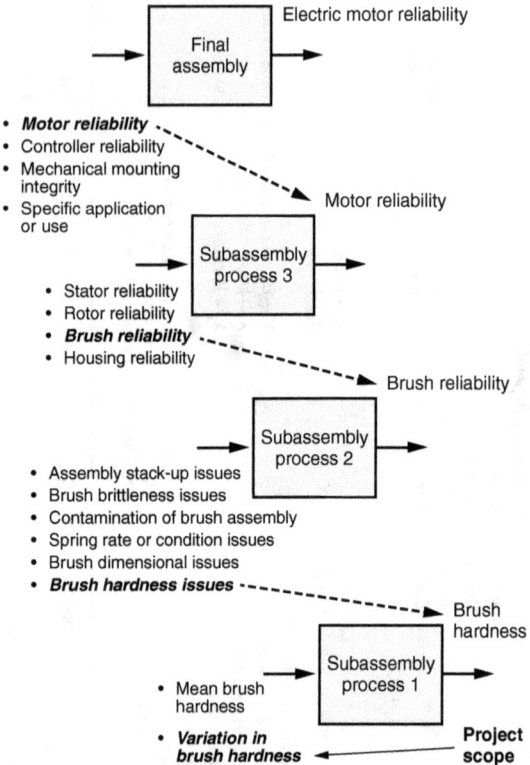

Figure 5.4   Example of project scoping using the dimensional method—process mapping decomposition format.

Chapter 6

Determining Project Savings

DEFINING HARD VERSUS SOFT DOLLARS

The discussion of what constitutes hard and soft dollar savings is a long-standing debate. It would seem to be a simple issue, but so many organizations have defined them differently over the years that confusion abounds. Ashenbaum (2006) noted, "Generally speaking, cost savings are understood as tangible bottom-line reductions resulting in saved money that could be removed from budgets or reinvested back into the business. There also needs to be a prior baseline or standard cost for the purchased product or service so that these savings could be measured against a prior time period's spend." (Note: In the context in which Ashenbaum offered this definition, cost savings are synonymous with hard dollar savings.) Based on this definition, let's extract the key characteristics of *cost (hard) dollar savings*:

- A prior baseline of spending must be established. This is typically considered 12 months.

- The dollars must have been planned and in the budget. This goes along with the baseline requirement above.

- The cost savings must affect the bottom line (that is, the profit and loss statement or balance sheet).

This definition also settles the question regarding whether spent or reinvested cost savings were ever savings at all. Some organizations tend to empty the savings pot as soon as it's filled. Then they wonder or even complain about the lack of savings from the Lean Six Sigma strategic initiative. Once a project books its hard savings, these savings exist regardless of what the organization chooses to do with those savings thereafter.

So, what are soft dollar savings? *Soft dollar savings* are cost avoidance savings and are, by exclusion, everything that is not hard dollar savings. This concept is illustrated in Figure 6.1. This figure makes the simple point that if it isn't hard dollar savings, then it is soft dollar savings. However, Ashenbaum offers some additional clarity on the subject:

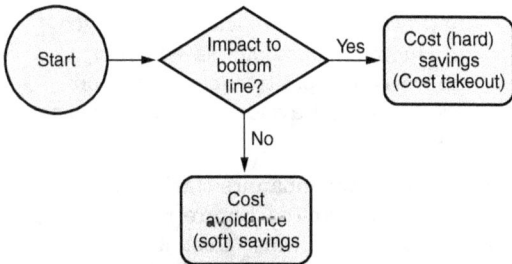

Figure 6.1 Determining hard dollar versus soft dollar cost savings.

- "Avoidance is a cost reduction that does not lower the cost of products/services when compared against historical results, but rather minimizes or avoids entirely the negative impact to the bottom line that a price increase would have caused.

- When there is an increase in output/capacity without increasing resource expenditure, in general, the cost avoidance savings are the amount that would have been spent to handle the increased volume/output.

- Avoidances include process improvements that do not immediately reduce costs or assets but provide benefits through improved process efficiency, employee productivity, customer satisfaction, improved competitiveness, and so on; over time, cost avoidance often becomes cost savings."

If we consider Ashenbaum's clarifications of soft dollars, we can expand Figure 6.1, resulting in Figure 6.2. Notice that "Cost avoidance" now depends on whether it has a budget impact or not. Let's consider the following examples:

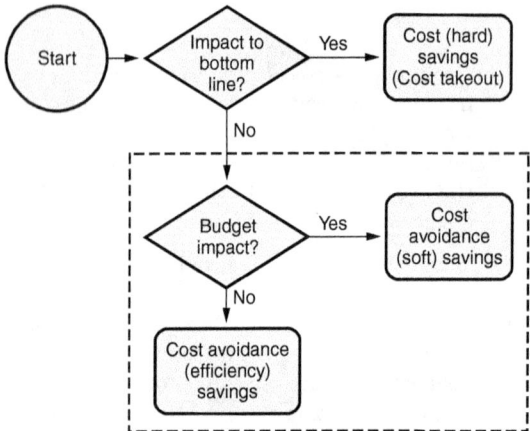

Figure 6.2 Expanding the definition of cost avoidance to include efficiency savings.

- *Cost avoidance (budget impacting).*
  This type of cost avoidance eliminates
  or reduces items in the budget marked
  for future spending. For example, three
  engineers are budgeted for hire in the
  fourth quarter. The cost of the three
  engineers is not in the baseline, nor
  has any spending for these engineers
  occurred. Eliminating these planned
  expenses has an impact on the future
  budget and is thus cost avoidance.

- *Cost avoidance (non–budget impacting).*
  This type of cost avoidance results from
  productivity or efficiencies gained in a
  process (that is, reduction of non-value-
  added activities) without a head count
  reduction. For example, the process cycle
  time that involved two workers was
  reduced by 10 percent. Assuming the
  10 percent amounts to two hours per
  worker per week, the two workers save
  four hours per week. These four hours
  are allocated to other tasks.

Another way to impact the bottom line is through
revenue or top-line growth. How might a Lean
Six Sigma project affect such growth? Consider a
project that identifies and removes a production
bottleneck or constraint. As a direct consequence
of removing that constraint, the organization is

able to accept new customer orders (something it wasn't able to do previously) or work off back-logged orders.

In this situation, revenue growth is achieved, and the reason is the constraint removal as an outcome of the project. It would seem that "revenue growth" would be a desirable category to which to attribute the dollar impact of Lean Six Sigma project dollars. This is illustrated in Figure 6.3.

Some organizations further refine cost savings into the third category of working capital/cash flow. Projects falling into this category might improve the internal invoicing process, resulting in lower accounts receivables, reduced inventory requirements through increased process efficiency, and reduced claims. Working capital/cash flow is reflected in Figure 6.4.

In summary, two major categories of savings have been identified, each of which can be further refined:

- Hard dollar savings (cost savings)

  - Cost takeout

  - Revenue growth

  - Working capital/cash flow

- Soft dollar savings (cost avoidance)

  - Cost avoidance (budget impacting)

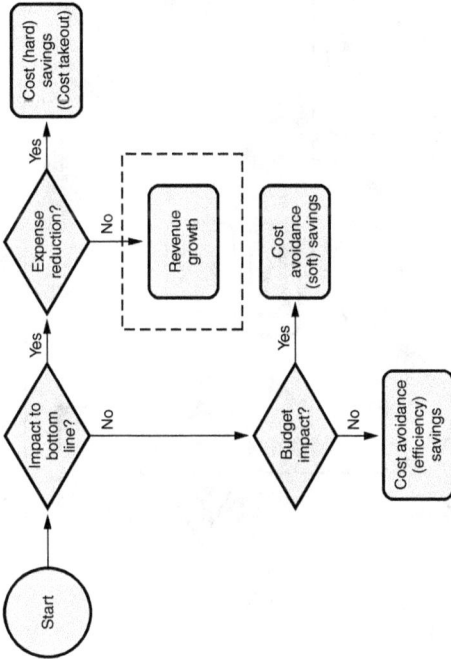

Figure 6.3 Expanding the definition of cost savings to include revenue growth.

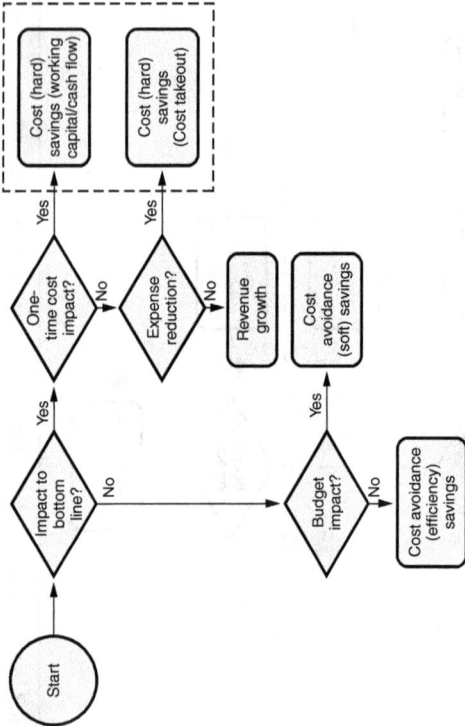

Figure 6.4 Expanding the definition of cost savings to include working capital.

   – Cost avoidance (non–budget
     impacting)

At one time or another, several of the progenitors
of Lean Six Sigma, such as AlliedSignal (later
Honeywell International) and GE, favored the
categories of savings similar to those above.

FALSE SAVINGS

One of the most troubling and abused areas of
savings is cost avoidance (non–budget impacting). This area of savings is typically generated
from productivity or efficiency gains. Many organizations will save a small slice of time in an
employee workday, multiply that by the number
of workers performing the same job, and extend
that over a year. Voilà! Enormous savings are
generated and booked.

   This approach is common in many transaction-based organizations such as large banks and
call centers, where a few seconds may be saved on
a transaction or in calling up a computer screen
of customer data. These transactions may be performed hundreds of thousands of times per day
across multiple individuals.

   On the surface, this approach seems fine and
is generally accepted. However, the fallacy lies
in the ability to aggregate these miniscule time

slices into a sufficiently meaningful size that will accommodate other work. If this can't be done, the "time saved" will simply be absorbed into adjacent work activities. Then they are not even paper savings.

Furthermore, the ability to measure such time savings might be questionable. Consider trying to measure five or ten seconds saved from a two- to three-minute transaction. In many cases, we are dealing at the noise level. Yet, these types of savings are reported all the time. Although they may be categorized as cost avoidance due to productivity or efficiency gains, are they really that if the gains can not actually be used or even measured reliably?

IDENTIFYING PROJECT SAVINGS

Table 6.1 provides a starting point for categorizing improvement in business metrics into the various categories of savings at the outset of a project. However, at the beginning of a project, our knowledge of potential savings is limited. Some categories of savings by business metric may hold throughout the project duration with only minor increases or decreases. Others that were initially projected may never materialize. Still others that were unforeseen may surface near the project end and yield healthy returns.

Table 6.1 Examples of savings by category.

Business metric	Savings categories				
	Cost (hard) savings			Cost (soft) avoidance	
	Cost takeout (bottom line)	Revenue growth (top line)	Working capital/cash flow	Cost avoidance (budget)	Cost avoidance (non-budget)
Labor					
The amount of labor required to perform a process decreases					
Shift premium decreases or is eliminated					
Overtime decreases					
Less rework occurs					
Material					
Less material is consumed (for example, solvents, office supplies)					
Less raw material is purchased					

Continued

Table 6.1 *Continued.*

Business metric	Savings categories					
	Cost (hard) savings			Cost (soft) avoidance		
	Cost takeout (bottom line)	Revenue growth (top line)	Working capital/cash flow	Cost avoidance (budget)	Cost avoidance (non-budget)	
Less material is stocked in inventory						
Shelf life is extended/expirations are reduced						
Reduced lead time in acquiring parts						
Reduced price of raw materials						
Increased inventory turns						
Inventory carrying costs are reduced						
Capacity						
Machine cycle time decreases						

Continued

Table 6.1 *Continued.*

Business metric	Savings categories					
	Cost (hard) savings			Cost (soft) avoidance		
	Cost takeout (bottom line)	Revenue growth (top line)	Working capital/cash flow	Cost avoidance (budget)	Cost avoidance (non-budget)	
Available capacity increases and can be used						
Overhead						
Travel expenses are reduced or eliminated						
Reduction of floor space						
Improvement in building and grounds						
Reduction in expenses and supplies						
Fewer utilities are used (for example, electric, gas, water)						

Continued

Table 6.1 *Continued.*

Business metric	Savings categories				
	Cost (hard) savings			Cost (soft) avoidance	
	Cost takeout (bottom line)	Revenue growth (top line)	Working capital/cash flow	Cost avoidance (budget)	Cost avoidance (non-budget)
Equipment					
Capital requirements are reduced or eliminated					
General and miscellaneous					
Less scrap is produced					
Customer penalties are decreased or eliminated					
Planned future costs are reduced or eliminated					
Less office supplies					
Customer					
Improvement in customer satisfaction					

Continued

Table 6.1 *Continued.*

Business metric	Savings categories					
	Cost (hard) savings				Cost (soft) avoidance	
	Cost takeout (bottom line)	Revenue growth (top line)	Working capital/cash flow		Cost avoidance (budget)	Cost avoidance (non-budget)
Improvement in voice of the customer						
Reduced warranty claims						
Increased OEM acceptance rates						
Process						
Increased yields						
Improvement in voice of the process						
Revenue growth						
Increased capacity						
Removal of process bottlenecks						

It is important to remember that any given project may have a wide variety of savings. Therefore, it is crucial to involve the financial representative to ensure that potential savings are not overlooked and to provide or enhance the credibility of the savings.

MULTIYEAR PROJECT SAVINGS

The recognized time frame for calculating a project's savings is 12 months. Consequently, we annualize a project's savings unless those savings had a one-time impact. Some organizations may argue that some projects legitimately save dollars over multiple years. However, savings in years two and beyond are generally built into budgets. Hence, they would reflect as a zero change as the previous 12 months serves as the baseline.

METHODS FOR DETERMINING COST BASELINES

The following are four common cost allocation models:

- *Undifferentiated costs*. This method considers those costs that are known or assumed associated with producing the product or rendering the service. There is no categorization by cause, type of product, or service.

- *Categorized costs (cost of quality)*. Such costs may be aligned to the cost of quality categories: appraisal, prevention, and failure.

- *Traditional cost model*. Costs allocated by a percentage of direct labor dollars, or some other arbitrary factor such as amount of space occupied, number of personnel in a work unit, and so on.

- *Activity-based costing (ABC)*. The costs of materials and services are allocated by the activity or process (cost drivers) by which they are consumed.

Each of the above models provides a method for establishing baseline performance costs for processes.

Chapter 7

Tools of the Trade

DMAIC

Tables 7.1 through 7.5 have been compiled to reflect the current thinking among contemporary authors about which tools should be applied in which phases of the DMAIC methodology. It should be readily apparent that these tables extend well beyond the alignment of tools by phase outlined in the ASQ Bodies of Knowledge for the Green Belt, Black Belt, and Master Black Belt certifications.

However, all of the tools are covered in the corresponding three handbooks: *The Certified Six Sigma Green Belt Handbook*, *The Certified Six Sigma Black Belt Handbook*, and *The Certified Six Sigma Master Black Belt Handbook*. As the reader and a practitioner of Lean Six Sigma, you may or may not agree with what the noted authors in the field of Lean Six Sigma

Table 7.1   Common tools used in *define* phase.

5 whys	*Data collection plan*	*Prioritization matrix*
Activity network diagrams	*Failure mode and effects analysis (FMEA)*	Process decision program chart
Advanced quality planning		Project charter
Affinity diagrams	*Flowchart/process mapping (as is)*	*Project management*
Auditing	Focus groups	Project scope
Benchmarking	*Force-field analysis*	*Project tracking*
Brainstorming	*Gantt chart*	Quality function deployment (QFD)
Cause-and-effect diagrams	*Interrelationship digraphs*	*Run charts*
Check sheets	Kano model	Sampling
Communication plan	Matrix diagrams	Stakeholder analysis
Control charts	*Meeting minutes*	*Supplier–input–process–output–customer (SIPOC)*
Critical-to-quality (CTQ) tree	*Multivoting*	Tollgate review
Customer feedback	*Nominal group technique*	*Tree diagrams*
Customer identification	*Pareto charts*	*Y = f(X)*
Customer interviews	*Project evaluation and review technique (PERT)*	
Data collection		

Table 7.2 Common tools used in *measure* phase.

Basic statistics	*Hypothesis testing*	*Project tracking*
Brainstorming	*Measurement systems analysis (MSA)*	Regression
Cause-and-effect diagrams		*Run charts*
	Meeting minutes	*Scatter diagrams*
Check sheets	Operational definitions	Spaghetti diagrams
Circle diagrams		*Statistical process control (SPC)*
Correlation	*Pareto charts*	
Data collection	Probability	*Supplier–input–process–output–customer (SIPOC)*
Data collection plan	*Process capability analysis*	
Failure mode and effects analysis (FMEA)		Taguchi loss function
	Process flow metrics	*Tollgate review*
Flowcharts	*Process maps*	*Value stream maps*
Graphical methods	*Process sigma*	
Histograms	*Project management*	

Table 7.3   Common tools used in *analyze* phase.

Affinity diagrams	*Histograms*	*Project tracking*
ANOVA	*Hypothesis testing*	Qualitative analysis
Basic statistics	*Interrelationship digraphs*	Regression
Brainstorming		
Cause-and-effect diagrams	Linear programming	Reliability modeling
Components of variation	Linear regression	Root cause analysis
Design of experiments (DOE)	Logistic regression	*Run charts*
Exponentially weighted moving average charts	*Meeting minutes*	*Scatter diagrams*
	Multi-vari studies	Shop audits
Failure mode and effects analysis (FMEA)	Multiple regression	*Simulation*
	Multivariate tools	*Supplier–input–process–output–customer (SIPOC)*
Force-field analysis	Nonparametric tests	
Gap analysis	Preventive maintenance	*Tollgate review*
General linear models (GLMs)	*Process capability analysis*	*Tree diagrams*
Geometric dimensioning and tolerancing (GD&T)	*Project management*	Waste analysis
		$Y = f(X)$

Table 7.4   Common tools used in *improve* phase.

Activity network diagrams	*Measurement systems analysis (MSA)*	*Project management*
Analysis of variance (ANOVA)	*Meeting minutes*	*Project tracking*
Brainstorming	Mixture experiments	Reliability analysis
Control charts	*Multi-vari studies*	Response surface methodology (RSM)
D-optimal designs	*Multivoting*	Risk analysis
Design of experiments (DOE)	*Nominal group technique*	*Simulation*
Evolutionary operations (EVOP)	*Pareto charts*	Statistical tolerancing
Failure mode and effects analysis (FMEA)	*Project evaluation and review technique (PERT)*	Taguchi designs
Fault tree analysis (FTA)	Pilot	Taguchi robustness concepts
Flowchart/process mapping (to be)	*Prioritization matrix*	*Tollgate review*
Gantt charts	*Process capability analysis*	*Value stream maps*
Histograms	*Process sigma*	Work breakdown structure
Hypothesis testing		$Y = f(X)$

Table 7.5 Common tools used in *control* phase.

5S	*Measurement systems analysis (MSA)*	Standard operating procedures (SOPs)
Basic statistics		
Communication plan	*Meeting minutes*	Standard work
Continuing process measurements	Mistake-proofing/poka-yoke	*Statistical process control (SPC)*
Control charts	Pre-control	*Tollgate review*
Control plan	*Project management*	Total productive maintenance
Data collection		
Data collection plan	*Project tracking*	Training plan deployment
	Run charts	
Kaizen	Six Sigma storyboard	Visual factory
Kanban		Work instructions
Lessons learned		

suggest for how the tools should be aligned, based on your personal experiences.

Nevertheless, a review of these tables indicates that certain tools have extensive use across the phases. For example, failure mode and effects analysis (FMEA) appears in four phases, thus indicating its importance to the DMAIC methodology and perhaps even the degree of emphasis that should be placed on it during training. For the reader's convenience, tools with applications in multiple phases have been italicized.

Another key aspect of these tables should be noted. It is obvious that many of the tools in these tables can be seen to fall under the general classification of other tools listed in the same table.

However, they are listed separately to allow visibility into the emphasis the various authors have placed on these tools. For example, *statistical process control* (SPC) is identified as a general classification of tools, while *control charts* is identified as a basic set of tools. Recall that the control chart is considered one of the seven basic quality tools, while SPC encompasses much broader concepts.

LEAN

As with Six Sigma, the tools of Lean are not new, only rediscovered. Table 7.6 lists many of the common tools used by Lean practitioners. Most of these tools are covered in the three ASQ handbooks previously mentioned.

While Six Sigma focuses on reducing process variation and enhancing process control, Lean—also known as lean manufacturing—drives out waste (non-value-added) and promotes work standardization and flow. These methodologies

Table 7.6 Common tools used in Lean.

5S	Line balancing	Spaghetti diagram/chart
Autonomation	Linear programming	Standard work
Batch/lot size reduction	Load leveling/level scheduling	Standardization
Brainstorming	Non-value-added activities analysis	Supermarkets
Buffer stock/inventory	Overall equipment effectiveness (OEE)	Takt time
Bullwhip effect		Theory of constraints
Changeover time	Poka-yoke	Throughput time
Continuous flow manufacturing (CFM)	Process efficiency	Total productive maintenance (TPM)
Cycle time	Product family	Transport time
Demand analysis/management	Pull systems	Value stream mapping (VSM)
Facility layout	Queue time	Visual control/management
First in, first out (FIFO)	Runner	
	Safety stock/inventory	Waste/waste reduction/7 or 8 wastes
Just-in-time	Simplification	
Kaizen	Single-minute exchange of die (SMED)	Work cell design
Kanban		Work-in-process (WIP) inventory
Lead time	Single-piece flow	

are complementary, not contradictory as some organizations and practitioners of each believe and defend.

In recent years, the demarcation between Six Sigma and Lean has blurred. We are hearing about terms such as "Lean Six Sigma" with greater frequency because process improvement requires aspects of both approaches to attain as well as sustain positive results. It is for this reason that Six Sigma practitioners need to be well versed in the tools of Lean.

Many successful implementations have begun with the lean approach, making the workplace as efficient and effective as possible, reducing the (now) eight wastes, and using value stream maps to improve understanding and throughput. When process problems remain, the more technical Six Sigma statistical tools may be applied. Furthermore, both methodologies require strong management support to make them the standard way of doing business.

Some organizations have responded to this dichotomy of approaches by forming a Lean Six Sigma problem-solving team with specialists in the various aspects of each discipline, but with each member cognizant of others' fields. Task forces from this team are formed and reshaped depending on the problem at hand.

Unfortunately, other organizations have decided to adopt one approach to the exclusion of the other. Perhaps this is due to a strong influence by uninformed senior management coupled with both weak skills and leadership on the part of the deployment leader.

Regardless, Black Belts must be evangelists and passionate about Lean Six Sigma. They must be able to carry forth the word and communicate the symbiosis that exists between Lean and Six Sigma, or any other methodology for that matter, and how those relationships can be leveraged to the benefit of the organization. They will be viewed as the expert. They must be out front on the firing line ready to convert the naysayers.

Chapter 8

Lean Six Sigma Roles and Responsibilities

INTRODUCTION

Enterprises with successful Lean Six Sigma programs have found it useful to delineate roles and responsibilities for various people involved in project activity. Leaders must have full understanding and buy-in, otherwise failure is assured. Although titles vary somewhat from organization to organization, the list below represents the general thinking regarding each role.

EXECUTIVE MANAGEMENT

The definition and role of leadership have undergone major shifts over the years. The leadership model that is most effective in the deployment of Lean Six Sigma envisions the leader as a problem solver. The leader's job is to implement systems that identify and solve problems that impede the effectiveness of processes.

This concept of leadership implies an understanding of team dynamics and Lean Six Sigma problem-solving techniques. Deployment of any culture-changing initiative such as the adoption of Lean Six Sigma rarely succeeds without management support. Initiatives starting in the rank-and-file seldom gain the critical mass necessary to sustain and fuel their own existence.

Successful implementations of Lean Six Sigma have unwavering support from the company-level executives (also known as *top management*). Therefore, it is imperative that executive management be involved in at least the following activities:

- Demonstrating support for the initiative by serving as a role model and exhibiting positive behaviors

- Communicating the importance of the initiative

- Setting the vision and overall objectives for the initiative

- Setting the direction and priorities for the Lean Six Sigma organization

- Approving the project selection methodology

- Selecting/suggesting projects

- Ensuring projects are aligned to organizational strategy

- Selecting project champions

- Allocating resources for projects

- Monitoring the progress of the initiative

- Modifying compensation systems (may reward successful projects)

- Establishing new policies related to or brought about by Lean Six Sigma

Executives typically receive specialized Lean Six Sigma training to support their role.

Occasionally, it may be necessary to assess the level of support received from executive management. Figure 8.1 provides an example of a radar chart that identifies several dimensions of management support. Criteria for scoring each dimension can be developed easily to accommodate most organizations' needs. Where necessary, plans may be put in place to fortify those dimensions that require strengthening.

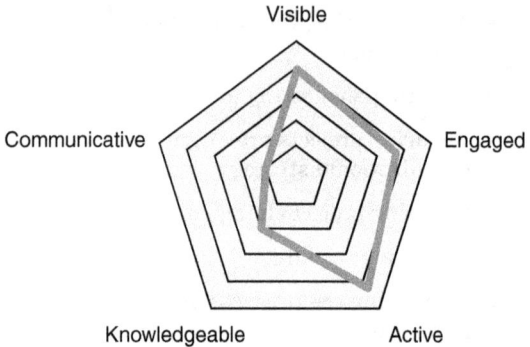

Figure 8.1 Dimensions of management support.

CHAMPION (SPONSOR)

A *champion* is a Lean Six Sigma role associated with a senior manager (usually an executive) who:

- Approves and signs off on the project charter

- Assumes ultimate responsibility for the success of the project

- Serves as the liaison with executive management

- Serves as the team's backer

- Ensures his or her projects are aligned with the organization's strategic goals and objectives

- Terminates projects that no longer align with strategic goals and objectives

- Provides the Lean Six Sigma team with resources

- Removes organizational barriers for the team

- Leads project tollgate reviews

- Asks appropriate questions of the project team

- Approves completed projects

- Provides reward and recognition to project team members

- Leverages (replicates) project results

Champions typically receive specialized Lean Six Sigma training to support their role.

A champion is also known as a *sponsor*. (*Note*: Some authors make a distinction between a champion and a sponsor. They consider the champion as the role defined here. However, they consider a *sponsor* to be an executive who assumes the roles and responsibilities identified with executive management above.)

PROCESS OWNER

A *process owner* is a Lean Six Sigma role associated with an individual who coordinates the various functions and work activities at all levels of a process, has the authority or ability to make changes in the process as required, and manages the entire process cycle so as to ensure performance efficiency and effectiveness. Process owners should be sufficiently high in the organization (ideally an executive) to make decisions regarding process changes. In many cases, the process owner is the same as the project champion.

It is only natural that managers responsible for a particular process frequently have a stake in keeping things as they are. They should be involved in any discussion of change. In most cases, they are willing to support changes but need to see evidence that recommended improvements are for the long-term good of the enterprise.

A team member with a "show me" attitude can make a very positive contribution to the team. Process owners should be provided with opportunities for training at least to the Green Belt level.

Specific roles and responsibilities of the process owner include:

- Provides process knowledge

- Reviews process changes

- Implements process changes

- Assumes responsibility for the execution and maintenance of the control plan

- Ensures improvements are sustained

In some cases, when the process is a large, cross-functional business process, the process owner may have lower-level process owners within each functional organization reporting indirectly to him or her. Consider the business processes shown in Figure 8.2. In these instances, the process owner will have the additional responsibility of providing coordination among the lower level process owners.

MASTER BLACK BELT

A *Master Black Belt* (MBB) is a Lean Six Sigma role associated with an individual typically assigned full-time to train and coach Black Belts to ensure that improvement projects chartered are the right strategic projects for the organization. Master Black Belts are usually the authorizing body to certify Black Belts and Green Belts.

Master Black Belts have advanced knowledge in statistics and other fields and provide technical support to the Black Belts and Green Belts.

Kubiak (2004) suggests a set of skills Black Belts must develop before they can become Master Black Belts:

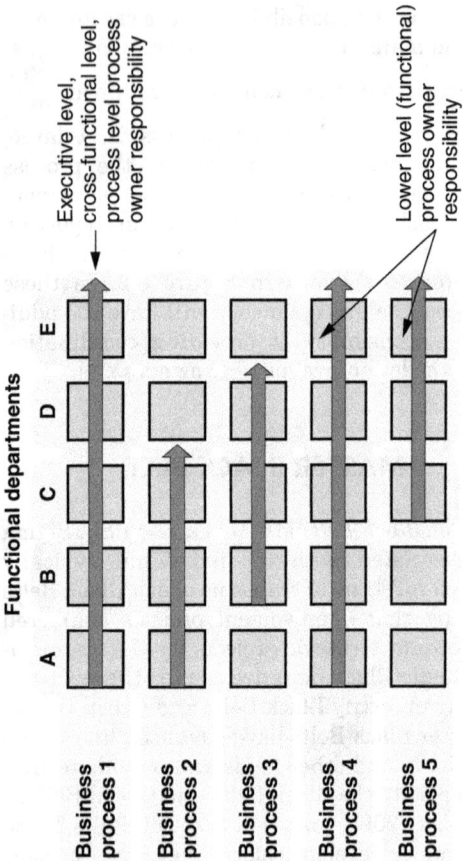

Figure 8.2 Cross-functional business processes.

- *Trainer.* Transfer of knowledge in a smooth and proficient manner is necessary in order for Black Belts to learn and conduct projects both efficiently and effectively.

- *Selector and scoper of projects.* Initiating meaningful and well-scoped projects will help complete them with minimal pain and suffering and will often achieve the anticipated results.

- *Coach.* This is a critical role of the Master Black Belt and a major contributor to the completion of projects.

- *Barrier breaker.* They must be able to break barriers for themselves instead of always relying on the champion.

- *Evangelist.* They must be passionate about Lean Six Sigma and carry forth the word, as they will be seen as the expert. They must be out on the firing line converting the naysayers.

- *Doer.* They must be able to roll up their sleeves and conduct projects from start to finish.

Master Black Belt projects may be of equal or greater complexity to those undertaken by the Black Belt. This depends, of course, on the total role of the Master Black Belt. In addition to projects, the MBB may be conducting training, consulting, coaching, and training development.

BLACK BELT

A *Black Belt* (BB) is a Lean Six Sigma role associated with an individual typically assigned full-time to train and coach Green Belts as well as lead improvement projects using specified methodologies such as define, measure, analyze, improve, and control (DMAIC) and Design for Six Sigma (DFSS), and other methodologies such as those shown in Figure 3.1.

The minimum size of a Black Belt project may be set by the organization. Regardless, it must be well defined by the charter document. The typical underlying process may cross multiple organizations and involve several key stakeholders.

Notice that this is defined as a full-time role. Generally, it has been found that part-time Black Belt roles have not been successful.

GREEN BELT

A *Green Belt* (GB) is a Lean Six Sigma role associated with an individual who retains his or her regular position within the organization but is trained in the tools, methods, and skills necessary to conduct Lean Six Sigma improvement projects either individually or as part of a larger team.

On a team, a Green Belt works under the direction of a Black Belt, providing assistance with all phases of project operation, while individually the Green Belt will receive coaching from a Black Belt or Master Black Belt. Generally, Green Belts are less adept at statistics and other problem-solving techniques.

Green Belt projects are typically small in scope and duration, and often confined to the GB's home functional department.

QUALITY COUNCIL

Bertin and Watson (2007) state that "Governance establishes the policy framework within which business leaders will make strategic decisions to fulfill the organizational purpose as well as the tactical actions that they take at the level of operational management to deploy and execute the organization's guiding policy and strategic direction." This, of course, refers to corporate governance in general.

Governance, in the context of Lean Six Sigma deployment, includes all the processes, procedures, rules, roles, and responsibilities associated with the strategic, tactical, and operational deployment of Lean Six Sigma. It also includes the authoritative, decision-making body charged with the responsibility of providing the required governance.

The concept of governance is critical to ensure a successful deployment. Many organizations omit this crucial component and wonder why the deployment stalls or even fails.

A common governance structure within a simple organizational unit is illustrated in Figure 8.3. Note: The Lean Six Sigma deployment team, led by the Lean Six Sigma deployment leader, reports indirectly to the governing body. This

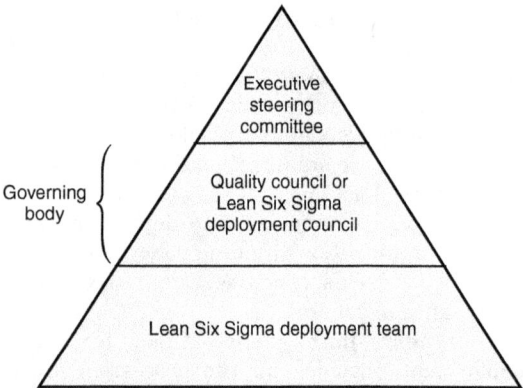

Figure 8.3 A common governance structure within a single organizational unit.

body is known by many names, commonly the *Quality Council* or *Lean Six Sigma Deployment Council.*

In addition, it is customary that the quality council reports locally within the organizational unit it serves to an *Executive Steering Committee.* Such a committee usually comprises the chief executive of the unit and his or her direct reports. The value of this committee is to engage the executive level, build ownership, and drive accountability.

Occasionally, the quality council will operate within a complex organizational structure such as the one shown in Figure 8.4. In this example, each chairperson of a tier 3 council would serve on his or her respective tier 2 council. Similarly, each chairperson of a tier 2 council would serve on the top or tier 1 council. This approach facilitates both vertical and horizontal communication throughout the entire organization.

Consider the various roles of a typical quality council as depicted in Figure 8.5. Let's briefly discuss each of these roles:

- *Create policies.* Policies provide an overarching framework for the deployment and set the governing principles of how the deployment will function. An example of a policy might be that the organization will only select projects that provide hard dollar bottom-line savings. All

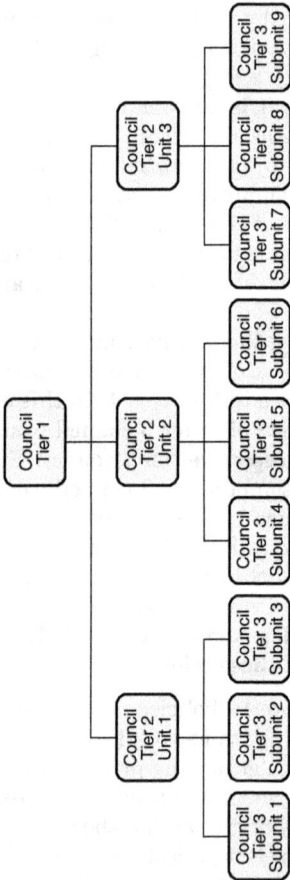

Figure 8.4 Nested governance structures within a complex organizational unit.

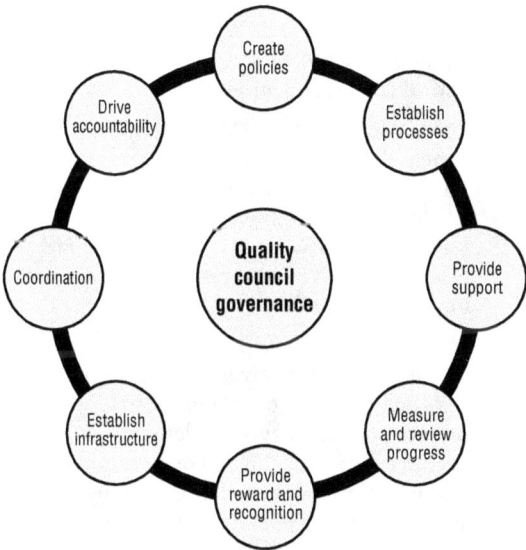

Figure 8.5 The many roles of a Lean Six Sigma Quality Council.

other categories of savings will not be considered.

- *Establish processes.* An example of this role might be the "project selection process" that identifies who selects the projects, what happens after they're selected, and by whom, and so on. While the quality council may not be the generator of the actual process document, it will have final authority over its approval and issuance.

- *Provide support.* Since the quality council has visibility into the overall deployment, it is in a position to work within the total organizational structure to bring resources to bear and reallocate resources among projects.

- *Measure and review progress.* The quality council receives progress reports regularly on project activities, attends project portfolio reviews, determines the successfulness of the deployment, and makes adjustments to the deployment strategy as appropriate.

- *Provide reward and recognition.* The quality council provides reward and recognition in a unified and appropriate manner per its policy. Many organizations fail to set such a policy, thus creating great inequities in delivering reward and recognition or failing to deliver any at all.

- *Establish infrastructure.* Infrastructure is often overlooked by many organizations. If not overlooked, it may be overtly avoided since infrastructure creates costs. Examples of infrastructure might be the creation of formal job descriptions for all employees of the Lean Six Sigma deployment team or the purchase of statistical analysis software. In some organizations, the infrastructure is considered the team members themselves.

What does this mean? It means that an organization desiring to deploy Lean Six Sigma might hire a Lean Six Sigma deployment leader but choose not to provide staff to the leader (that is, the people infrastructure necessary for the deployment is deliberately omitted). Organizations doing this are usually testing Lean Six Sigma, and hence leadership is not fully committed.

- *Coordination.* The role of coordination for a quality council may come in different forms. If the organization is small or simple in structure, the quality council may be required to coordinate projects across departments or even geographies. However, for complex organizations, this coordination role can be significantly compounded since there are many quality councils that it must work with laterally, those that report to it, and those to which it reports. See Figure 8.4.

- *Drive accountability.* Because quality councils in Lean Six Sigma deployments typically comprise member representatives across the breadth of the organization, they are in a strong position to drive accountability, a key component in any successful deployment. Some councils choose to make use of the responsibility, accountability, consultation, inform (RACI) matrix shown in Table 9.5 (see Price and Works, "Balancing Roles and Responsibilities in Lean Six Sigma").

If managed properly and staffed with strong leaders, the quality council can be a highly effective tool for driving a Lean Six Sigma deployment.

Chapter 9

Stakeholder Engagement and Communications

STAKEHOLDER ENGAGEMENT

Tague (2005) defines a *stakeholder* as "anyone with an interest or right in an issue, or anyone who can affect or be affected by an action or change. Stakeholders may be individuals, groups, internal or external to the organization." To help identify stakeholders related to a project, she recommends asking the following set of eight questions:

- Who might receive benefits?

- Who might experience negative effects?

- Who might have to change behavior?

- Who has goals that align with these goals?

- Who has goals that conflict with these goals?

- Who has responsibility for action or decision?

- Who has resources or skills that are important to this issue?

- Who has expectations for this issue or action?

A form similar to the one depicted in Table 9.1 can be used to capture relevant information about the stakeholders. Let's review the form:

- *Stakeholder.* The list provided above is an excellent starting point for identifying stakeholders. There are two types of stakeholders: primary and secondary. *Primary stakeholders* are directly affected by the project. *Secondary stakeholders* are involved in implementing, funding, monitoring, and so on, and are considered intermediaries. Consider identifying each stakeholder as primary or secondary.

- *Project impact.* Clarify the impact of the project on the stakeholder. This establishes the relationship between the stakeholder and the project.

- *Level of influence.* Rate the level of influence the stakeholder has on the project. Consider a scale from 1 to 5,

Table 9.1 A simple form for completing a stakeholder analysis.

Stakeholder	Project impact	Level of influence	Level of importance	Current attitude	Action plan

or low, medium, and high. Low and high
works well, too.

- *Level of importance.* Rate the level of
  importance each stakeholder has in the
  project. Consider a scale from 1 to 5, or
  low, medium, and high. Low and high
  works well, too.

- *Current attitude.* Rate the attitude the
  stakeholder has toward the project.
  Consider a scale from 1 to 5, or low,
  medium, and high. Low and high works
  well, too.

- *Action plan.* Outline different strategies
  for dealing with each stakeholder. The
  strategies should focus on reducing
  opposition and increasing support.

An alternate stakeholder analysis form is shown
in Figure 9.1. Notice that this form allows for the
categorization of stakeholders.

Once Table 9.1 has been completed, the
ratings from the level of influence and importance columns can be transferred to Table 9.2.
Table 9.2 provides the added benefit of classifying stakeholders:

- *High influence/high importance.*
  Collaborate with these stakeholders.

Stakeholder Analysis

Prepared By:					Date:			

Project:
Project

Description:
Use to organize a quality improvement project into phases that you define.

⑦ How to Use this Table

Stakeholder Categories	Relevant Stakeholders	Code	Attitude	Activity	Attitude Rating	Power	Interest	Power Rating
			0 ▾	0 ▾	0.00	0 ▾	0 ▾	0.00

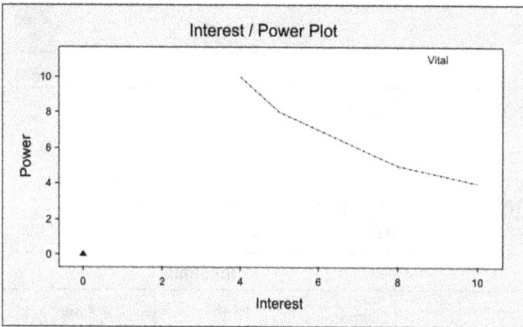

Figure 9.1   Alternate example of a stakeholder analysis.

Source: Courtesy of Minitab Quality Companion 3 software, Minitab Inc.

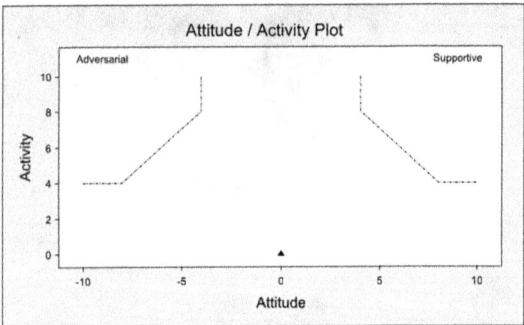

Figure 9.1　*Continued.*

Table 9.2　Example of an influence–importance stakeholder table.

		Influence	
		High	**Low**
Importance	**High**	Collaborate	Protect and defend
	Low	High risk	Do not spend resources

Source: Adapted from Tague (2005).

- *High influence/low importance.* Work with these stakeholders to involve them and increase their level of interest. These stakeholders are capable of sabotaging plans and escalating problems and are considered high risk.

- *Low influence/high importance.* Protect and defend these stakeholders. Work to give them a voice and help increase their level of influence.

- *Low influence/low importance.* Monitor these stakeholders, but don't spend resources on them.

Whenever possible, involve stakeholders with high influence; otherwise, neutralize their influence.

Andersen (2007) suggests another means of classifying stakeholders. His method provides additional insight and is depicted in Table 9.3. In this table, stakeholders are classified as to their potential impact on the organization versus their potential for cooperation with the organization. Again, this requires categorizing each stakeholder as either high or low. Once complete, stakeholders can be charted and viewed as:

- *High impact/high cooperation.* These stakeholders are considered a "mixed

Table 9.3   Example of an impact–cooperation stakeholder table.

		Potential to impact	
		High	**Low**
Potential for cooperation	**High**	Mixed blessing	Supportive
	Low	Nonsupportive	Marginal

Source: Adapted from Andersen (2007).

blessing" and should be handled through cooperation.

- *High impact/low cooperation*. These individuals are considered "nonsupportive." Minimize dependency on this type of stakeholder through a defensive strategy.

- *Low impact/high cooperation*. These stakeholders are considered "supportive" and should be involved in relevant discussions and decisions.

- *Low impact/low cooperation*. These stakeholders are considered "marginal" and should be monitored. Like Tague's low influence/low importance stakeholders, don't spend resources on them.

Black Belts may find a significant portion of their time dedicated to engaging project stakeholders. As Covey (1989) would say in his fifth habit, "Seek first to understand, then to be understood." The stakeholder analysis is a highly structured method dedicated to this purpose.

However, if done properly, it can be a very sensitive document. If left open to public consumption within the organization, it can be damaging, hurtful, or perceived as offensive. Use this tool effectively, but use it with discretion.

PROJECT STATUS COMMUNICATION

Ongoing communication throughout the duration of a project is essential. Two particularly useful documents for facilitating communication are the communications plan shown in Table 9.4

Table 9.4 Constructing a communications plan.

Who delivers	What message	To whom	By what method	How often

and the RACI matrix shown in Table 9.5. These documents help direct the flow of communication both within and outside of the team.

Communicating within the Team

Communication within the team should take place on a regular basis. All communication within the team should be open, honest, and with no fear of retribution. In general, team members should be aware of any project problems or issues before anyone outside of the team. No team member should ever be blindsided because project information bypassed him or her.

In addition, team members need to be apprised of their own performance. Too often, this valuable activity is ignored either because it is uncomfortable for the project leader or because human capital resources are limited and the project leader is reluctant to alienate a scarce resource.

Lack of adequate communication is one of the most frequently noted causes of team failure. Serious effort toward communication improvement should be made at each stage of team development.

In some situations, such as large projects or those with wide geographical boundaries, the development of a formal communications plan is necessary. This is particularly true in the case of virtual teams.

Table 9.5 Example of a RACI matrix.

Player \ Task	Set policy	Identify projects	Select projects	Work projects	Achieve results	Maintain gains	Coach projects
Executive	A	A	A		I		
Champion/ sponsor		A	R				
Process owner		C	I			A	
MBB	I	C		C	R		A
BB	I		I	R	R		A
GB			I		A		
Quality council	R	C	I		C		

Many teams operating in today's large organizations may never be face-to-face or, for that matter, ever see one another. Fortunately, technology facilitates communication.

However, careful planning for communication is still required. Accommodations for extreme time zone differences might be necessary, and technologies may be new to some sites and localities. Therefore, it is important to test any form of technical communication before operating it in a live environment.

With the advent of technology that permits round-the-clock, worldwide contact, the use of virtual teams has become a reality. As noted previously, virtual team members may never see one another. This makes the communication process that much more difficult.

If the communication process is conducted via audio only, team members have no ability to read body language. They are limited to verbal cues and voice intonations. Consequently, some team members may find it difficult to function as a virtual team member. Therefore, team communication must be carefully planned, monitored, and adjusted as necessary.

Communication outside the Team

As with communication within the team, communication outside the team should also take place on a regular basis. Such communications typically include status reports and presentations. Status reports should be brief. Generally, a bullet format is sufficient when grouped under a series of headings. Typical headings include:

- Accomplishments since last report
- Problems, issues, or concerns
- Actions being taken
- 30-day plans

A recipient of a status report in the above format might conclude that the project leader is indeed making progress, has incurred problems but has recognized them (some leaders don't), has taken mitigating action (therefore, I do not need to intervene), and is proceeding forward once more.

The recipient reads the status report and completes it with a feeling of confidence in the project leader. Poorly written status reports invite criticism and occasionally unwarranted scrutiny and intrusion.

Presentations are another format commonly used to communicate outside the team and are usually delivered to higher levels of the organization. The project leader may request a spot on the sponsoring executive's staff meeting agenda or may be invited to attend.

While the above bullet format is acceptable, more detail is required. Remember, the presentation will be interactive and usually face-to-face. Therefore, messages need to be clear and crisp. There is no room for ambiguity.

Furthermore, the presenter should expect interruptions and questions out of order and not related to the intended delivery sequence. Hence, the presenter's motto is "Be prepared!" As with poorly written status reports, a poorly delivered presentation can be a disaster that makes recovery difficult and often brings unwanted help.

COMMUNICATIONS PLAN AND THE RACI MATRIX

A communications plan (see Table 9.4) defines who will deliver, what will be communicated, to whom, how often, and by what means. Communications could include status reports, presentations, newsletters, and so on. Methods may be either formal or informal. In all cases, though, the communications plan must reflect the needs of the audience.

Because quality councils in Lean Six Sigma deployments typically comprise member representatives from across the breadth of the organization, they are in a strong position to drive accountability, a key component in any successful deployment. Some quality councils choose to make use of the responsibility, accountability, consultation, and inform (RACI) matrix shown in Table 9.5 (Price and Works, "Balancing Roles and Responsibilities in Lean Six Sigma"):

- *Responsibility.* Individuals who actively participate in an activity.

- *Accountability.* The individual ultimately responsible for results.

- *Consultation.* Individuals who must be consulted before a decision is made.

- *Inform.* Individuals who must be informed of a decision because they are affected. These individuals do not need to take part in the decision-making process.

The RACI matrix is a particularly useful form of communication.

Chapter 10

The Tollgate Review

PROJECT REVIEWS

Executives should be actively engaged in the review of Lean Six Sigma projects. Champions, in particular, must be engaged in a specific set of reviews related to the projects for which they are responsible. Recall that the role of the champion was addressed in Chapter 8. This set of reviews is known as *tollgate reviews*.

TOLLGATE REVIEWS

Tollgates first appeared in Figure 2.2, which demonstrated how projects flow through the DMAIC process and the impact tollgate reviews have upon them. Tollgate reviews are a necessary part of the DMAIC infrastructure. They exist as a check and balance to ensure that teams are ready to transition from one phase to the next in an orderly manner. Tollgates are the bonds that hold the DMAIC framework together. They serve as meaningful checkpoints, facilitate communications, and ensure champion participation. Tollgates focus projects on moving forward.

The *tollgate review* is a formal review process conducted by a champion who asks a series of focused questions aimed at ensuring that the team has performed diligently during this phase. The result of a tollgate is a "go" or "no-go" decision. The go decision allows the team to move forward to the next phase. If it is in the last phase, the go decision brings about project closure. If the decision is no-go, the team must remain in the phase or retreat to an earlier phase, or perhaps the project is terminated or suspended. The concept of go/no-go decisions is illustrated in Figure 2.2.

Diligent application of the proper tools and, above all, significant preparation are required to pass tollgate reviews. Passing them is not a given. Notice that up to and including the analyze phase, the failure of a tollgate could set the project back one or two phases. This usually occurs when knowledge is gained in a phase contrary to expectations. In this case, the project team may be forced to retreat all the way back to *define* in order to restate the problem. Beyond the analyze phase, a failed tollgate usually results in correcting or performing additional work within the phase. However, a project may be terminated at any phase. Generally, this will happen when projects are no longer synchronized with strategy, champions/sponsors have lost interest or transferred, or the potential savings are less than expected.

Effective tollgate reviews are usually characterized by an objective evaluation of the work performed by the team during the phase and the willingness to identify and resolve problems during the reviews. Poor preparation, incomplete documentation, champions desiring to change the scope, and champions delegating replacements typically characterize poor tollgate reviews.

Many thoughts exist regarding the purpose of tollgates, who should attend them, what questions should be asked, and what the exit criteria from the tollgate phase are. Table 10.1 provides a summary of one viewpoint. The "questions to address" are minimal in the table, but they are critical. Notice that the first question asked is "Is this project still consistent with the goals of the organization?" Project alignment to the strategy, and subsequently the goals and objectives, of the organization is a critical focus of every tollgate review. This question appears in every phase because of the highly dynamic nature of organizations. When the answer is "no," it is incumbent on the champion to shut the project down and allow the project resources to be allocated elsewhere.

Table 10.1 A brief summary of the tollgate process.

Component	Purpose	Participants required	Questions to address	Exit criteria
Define	Provide a compelling business case appropriately scoped, complete with SMART goals, and linked to a hoshin kanri plan	• Sponsor • Process owner • MBB/BB coach • Deployment champion • Project team • Finance partner	• Is this project consistent with the goals of the organization? • Do we have the right level of engagement from stakeholders and business partners? • Have resources been allocated to move to the next phase? • Do conflicts exist with other projects or activities?	• Sponsor approval • Finance approval • Funding approval
Measure	Collect process performance data for primary and secondary metrics to gain insight and understanding into root causes and to establish performance baselines	• Sponsor • Process owner • MBB/BB coach • Deployment champion • Project team	• Is this project still consistent with the goals of the organization? • Do we have the right level of engagement from stakeholders and business partners? • Have resources been allocated to move to the next phase?	• Sponsor approval

Continued

Table 10.1 *Continued.*

Component	Purpose	Participants required	Questions to address	Exit criteria
Analyze	Analyze and establish optimal performance settings for each X and verify root causes	• Sponsor • Process owner • MBB/BB coach • Deployment champion • Project team • Finance partner	• Is this project still consistent with the goals of the organization? • Do we have the right level of engagement from stakeholders and business partners? • Have resources been a located to move to the next phase? • What are the market and timing dependencies?	• Sponsor approval • Finance approval
Improve	Identify and implement process improvement solutions	• Sponsor • Process owner • MBB/BB coach • Deployment champion • Project team	• Is this project still consistent with the goals of the organization? • Do we have the right level of engagement from stakeholders and business partners? • Have resources been allocated to move to the next phase?	• Sponsor approval

Continued

Table 10.1   *Continued.*

Component	Purpose	Participants required	Questions to address	Exit criteria
Control	Establish and deploy a control plan to ensure that gains in performance are maintained	• Sponsor • Process owner • MBB/BB coach • Deployment champion • Project team • Finance partner	• Is this project still consistent with the goals of the organization? • Have resources been allocated to move into replication? • Is the replication schedule and plan appropriate? • What are the market and timing dependencies? • Have responsibilities identified in the control plan been transferred to appropriate parties?	• Sponsor approval • Finance approval • Hand-off to process owner

Table 10.2 provides an additional set of questions the champion might ask during tollgate reviews. Of course, common to each phase are questions regarding budget and schedule. Also included is a question regarding what obstacles the team might have encountered and how the

Table 10.2 Additional questions to be asked by the champion during tollgate reviews.

Phase	Questions
Define	• Who are the stakeholders? • What are the primary and secondary measures? • Are you facing any barriers or obstacles I can help with? • Are we on schedule/budget?
Measure	• What is the process capability? • What data still remain to be collected? • What did the FMEA show? • What questions still remain to be answered about the process? • Are you facing any barriers or obstacles I can help with? • Are we on schedule/budget?
Analyze	• What are the critical X's? • What are the root causes? • Are you facing any barriers or obstacles I can help with? • Are we on schedule/budget?

Continued

Table 10.2 *Continued.*

Phase	Questions
Improve	• What is the process capability of the revised process? • What criteria will be used to select the optimum solution from the set of solutions? • Will we run a pilot? • Are you facing any barriers or obstacles I can help with? • Are we on schedule/budget?
Control	• Is the process documented? • Is the control plan completed? • Has the process been transitioned back to the process owner? • What lessons have we learned? • Do we know where we might be able to replicate our new process across the organization? • Are you facing any barriers or obstacles I can help with? • Are we on schedule/budget?

Source: Adapted and compiled from Phillips and Stone (2002).

champion might be able to help. Recall that one job of the champion is the removal of organizational barriers the team might face.

Tollgate reviews, when executed well, provide real value to the Lean Six Sigma deployment. They keep projects focused and on track, and executives and champions engaged.

Chapter 11

Coaching and Mentoring

COACHING

Coaching is a process by which a more experienced individual helps enhance the existing skills and capabilities that reside in a less experienced individual. Coaching is about listening, observing, and providing constructive, practical, and meaningful feedback.

During training, coaching helps the trainee translate the theoretical learning into applied learning while helping the trainee develop confidence in their newly obtained knowledge and skills. Post training, coaches help projects stay on track and advance toward completion in a timely manner.

Coaches provide guidance and direction for Belts on how to navigate organizational barriers; select and use the proper tools and techniques; prepare for tollgate reviews, project reviews, or other presentations; discover solutions on their own; provide intervention where needed; and generally serve as a sounding board or go-to person for project-related issues.

One key point about coaching can not be emphasized enough. Coaches rarely provide absolute direction. Instead, they help Belts learn on their own by extricating and synthesizing education, training, and past experiences through effective listening and carefully placed questions.

In some cases, the coach may find that the individual is unsuited to work a specific project and consequently will have the project assigned to another Belt or request that the project champion have the project canceled if it no longer appears viable.

EVALUATING THE TEAM'S PERFORMANCE

At various points in the life of a team, the coach will want to evaluate a Green Belt team's performance. The measurement system and criteria should be agreed on in advance. The criteria against which the team is evaluated must relate to progress toward goals and objectives. It is common practice to compare team progress against the timeline set forth in the project schedule.

Typical objective-oriented criteria include measurement against:

- Goals and objectives (to determine if the project is succeeding or failing)
- Schedule
- Budget

In addition to the above, it may be possible for the team coach to evaluate the team on other criteria, some of which are less objective:

- Attitudes
- Teamwork
- Attendance
- Following team norms
- Length in team stages (that is, forming, storming, norming, performing)

The coach will want to attend team meetings from time to time to make evaluations and observations. Above all, the coach will want to determine whether the team is making progress or is heading for a state of imminent failure.

If failure is close or imminent, the coach should take immediate action to prevent it. This form of action is called *intervention*. An intervention might take the form of replacing the team leader, replacing team members, adding a professional facilitator to the team, providing additional training to specific members, or whatever the coach decides is appropriate.

Regardless, the coach should frequently discuss his or her evaluations and observations with the team, as well as follow up with additional evaluations and observations at later meetings to determine whether behavior changes have occurred.

COACHING NON-BELTS

At the outset of most Lean Six Sigma deployments, organizations offer general awareness courses to their employees to explain "what this Lean Six Sigma stuff is all about." This training usually satisfies the needs of most employees, who never see themselves as becoming Lean Six Sigma Belts.

Over and above the general awareness training, most, if not all, Lean Six Sigma deployment departments develop their own intranet site. These sites provide additional information about courses, criteria for becoming Belts, expectations of Belts, status of in-process projects, completed projects, lists of sponsors, and so on. On some sites, the actual training material is visible to all employees, thus providing them with an opportunity to make an informed decision, or a foundation for a meaningful discussion with a Black Belt or Master Black Belt, about making Lean Six Sigma a career choice.

As with any organization, attrition occurs. If the population of Belt-level employees is not replenished as they move up or out into the organization, the Lean Six Sigma initiative will eventually grind to a halt. Therefore, it is imperative that the organization maintains an ongoing recruitment program for future Belts.

Several ways some organizations have found effective for recruiting future Belts include holding brown bag lunches, road shows (that is, presentations at department staff meetings), and targeted one-on-one meetings. These methods work well for the general population of employees.

In organizations where leaders are exceptionally committed and high profile, they may actively seek a coaching relationship with a Black Belt or Master Black Belt to work their Green Belt or Black Belt training into their busy schedules. Other organizations have proactively established executive training programs to accomplish the same thing. Executives are paired with Black Belts and Master Black Belts. Training is conducted one-on-one, and projects are still expected.

COACHING AS A COST ELEMENT

Coaching is a cost element some organizations like to omit during deployment, particularly if it is seen as an external cost and not viewed as value-added by management.

In fact, coaching adds significant value to the deployment. It helps expedite the assimilation and understanding of knowledge gained in the classroom by transferring theory into application. Furthermore, the mere presence of a coach who meets regularly with a candidate serves to accelerate projects to successful completion.

As with training, coaching may be conducted by internal or external resources. Although there is great disparity among coaches, a typical recommendation for coaching during Belt training is approximately one hour per week between classes. Some coaches vary this recommendation according to the belt level or the skill level of the individual Belt.

Use of internal coaches generally does not incur additional costs to the organization, but it does commit and constrain resources. As you plan for coaching, consider who will do the actual coaching, and the associated time and cost of each coach.

Common coaching schemes in use are:

- Certified Black Belts coach Green Belt candidates.

- Master Black Belt candidates coach
  Green Belt candidates.

- Master Black Belt candidates coach
  Black Belt candidates. This approach
  is used to help Master Black Belt
  candidates develop strong coaching
  skills, which will be necessary for
  coaching executives, champions, and
  sponsors.

- Certified Master Black Belts coach
  Master Black Belt candidates, Black
  Belt candidates, and Green Belt
  candidates.

External coaching resources can represent a
sizeable cost to the organization. However, they
are additional resources, thus freeing Black
Belts and Master Black Belts within the organization for project work.

Regardless of whether your organization
chooses to conduct coaching with internal, external, or a mix of resources, a cost–benefit analysis
is required to make an informed decision.

MENTORING

While coaching focuses on the individual as they
relate to Lean Six Sigma, *mentoring* focuses on
the individual from the career perspective.

Mentors are usually experienced individuals (not necessarily in Lean Six Sigma) who have in-depth knowledge about the organization as well as the individual (that is, mentee). Usually, they come from within the organization, though not necessarily from the same department as their mentee.

Their role is to help provide guidance, wisdom, and a possible road map to career advancement. Like the coach, the mentor also helps the individual navigate the organization. It would be conceivable that both a coach and a mentor would support an individual Belt concurrently. The coach would focus on the individual's role as a Belt, and the mentor would focus on the individual's overall career.

Chapter 12

Process Mapping

INTRODUCTION

Process maps reveal much more about a process than flowcharts do, and they provide the Black Belt with more guidance. Process maps address the key foundational concept of Lean Six Sigma: the concept of $Y = f(X)$, or simply, outputs are a function of inputs.

However, if the Lean Six Sigma project should take on more of a lean focus, the process map should evolve into a *value stream map* (VSM). The value stream map differs from the process map in that data elements relevant to lean are included. See *The Certified Six Sigma Black Belt*, Second Edition, for details.

CATEGORIZING AND CLASSIFYING PROCESS INPUT VARIABLES

Kubiak (2007) describes the basic anatomy of a process map, as shown in Figure 12.1. Input variables, our list of X's, flow into a transform function that we call a process. Flowing from the process is one or more desired outputs.

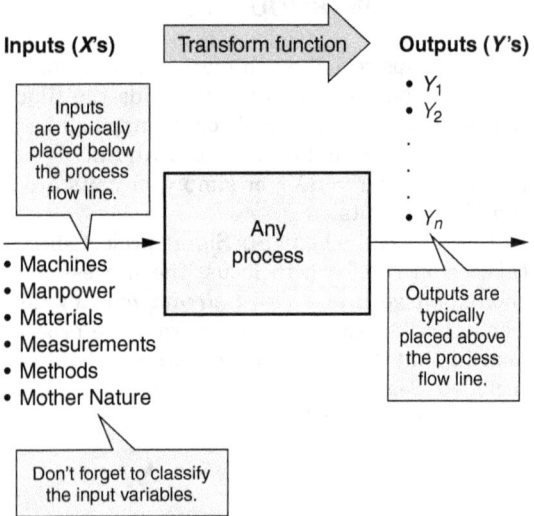

Inputs (X's) — Transform function — **Outputs (Y's)**

- Y_1
- Y_2
- Y_n

Inputs are typically placed below the process flow line.

- Machines
- Manpower
- Materials
- Measurements
- Methods
- Mother Nature

Any process

Outputs are typically placed above the process flow line.

Don't forget to classify the input variables.

Figure 12.1 Demonstrating the $Y = f(X)$ concept.

Notice that the inputs appear below the flow line, while outputs appear above the flow line. This is simply a matter of aesthetics.

Otherwise, inputs and outputs appearing on the same line would significantly clutter the map, particularly when multiple process blocks appear adjacent to one another.

Notice the list of input variables. They should be familiar as they represent the typical categories (also known as the six M's) of the main bones in a fishbone diagram. (Note: Some authors include "management" as another category and call the list the "seven M's."). Using the six M's as a structured approach provides some assurance that few inputs will be overlooked. Table 12.1 identifies each of the M's along with alternative terminology that may be used in service- or transaction-based industries, for example. Commentary and insight about each "M" have been included.

One important aspect regarding input variables not visible in Figure 12.1 is their classification. Classifying input variables helps Lean Six Sigma practitioners focus on those inputs that are controllable and guides practitioners away from spending time and energy on those that are not.

An example of a common classification scheme is shown in Table 12.2. Properly classifying input variables requires recognizing the need to fully

Table 12.1 The six M's: useful categories when thinking about input variables.

Input (X)	Alternate terminology	Comment
Machines	Equipment	Machines or equipment need not be costly or even high-tech. Don't overlook the basics. For example, service and/or transactional-based processes might include the use of simple devices such as a stapler or highlighter. When considering input variables of this nature, it is often helpful to consider developing a list of "tools required."
Manpower	People	Human resources may take various forms such as skilled technicians, engineers, or administrative and clerical personnel. Even highly automated processes may occasionally call for human support when preventive or corrective maintenance or actions are required.
Materials	Materials	Materials may include raw materials or even intermediate subassemblies. Materials are often consumed or transformed during the execution of a process.

Continued

Table 12.1   *Continued.*

Input (X)	Alternate terminology	Comment
Measurements	Measurements	Always remember to ensure the measurement system is capable. If a measurement system is assumed capable when, in fact, it is not, erroneous results may occur. These include overlooking true critical input variables or concluding some variables are critical when they are not.
Methods	Processes	Processes come in all shapes and sizes. They may be well defined or very loosely defined. In the context of a manufacturing environment, there is a category of processes known as "special processes."
Mother Nature	Environment	Variables falling into this category can be associated with either an internal or external environment. This is an important distinction, particularly when determining whether such variables are "noise" variables. For example, temperature and humidity would likely be considered controllable variables when the underlying process takes place in a clean room environment. However, if another process is conducted outdoors, temperature and humidity might very likely be considered noise variables since they may be impossible or too costly to control.

Table 12.2 Classifying input variables.

Variable designation	Type of variable	Comment
C	Controllable	These are variables over which the process owner has control (regardless of whether such control may ever be exercised). More specifically, characteristics or values of controllable variables can be set or manipulated in a manner that drives one or more output variables (Y's) in the desired direction.
N	Noise	Noise variables are those that either can not be controlled or perhaps are too expensive to control. It is important that such variables be defined so that the Green Belt team or project Black Belt knows which variables should not be addressed. Attempts to control "noise" variables often result in frustrated teams and/or failed projects.

Continued

Table 12.2 *Continued.*

Variable designation	Type of variable	Comment
SOP	Standard operating procedure	A standard operating procedure is a unique and predefined way of performing a process. For example, it may be an instruction document for assembling a bicycle or preparing an expense report. Just because an input variable is defined as an SOP, no inference should be drawn regarding the quality of the process encompassed by the procedure. SOP variables are a subset of controllable variables. Designating an input variable as SOP does not exclude it from the process owner's control. However, it does suggest that minimal variation is probably associated with it.
X	Critical	Critical input variables are a subset of the set of controllable variables. These are variables that have been determined to have a significant impact on one or more output variables (Y's). Significance may be demonstrated through statistical tools such as design of experiments, regression, etc. Early in the DMAIC phase and particularly before the completion of *analyze*, use of this designation should be considered, at best, tentative. Only after the analyze phase has been completed will we have the knowledge to state with some degree of certainty that a variable is a "critical input." Be careful not to confuse this classification of variable with the "X" used in the equation $Y = f(X)$.

understand a process in relation to the culture and business needs of the organization.

Let's assume that a technician represents the manpower required to perform a given process. What classification or variable designation from Table 12.2 should a Black Belt assign it?

- If the team determines that the skill and experience level of the technician can be assigned appropriately to the needs of the process, then the team would classify "technician" as a controllable input variable. It would appear in the process input list as "technician (C)." The input variable is followed by the classification designation in parentheses.

- If the team determines that management will never invest in training technicians and that the technician assigned to the process is fixed, the team may consider classifying "technician" as a noise input variable. If this were the case, then it would appear in the process input list as "technician (N)."

This example illustrates the need to revisit processes on a periodic basis since organizations continually change and evolve. New and insightful management may recognize the need to grow and improve the skill level of its technicians.

Hence, input variables previously classified as noise may now be considered controllable. Likewise, the opposite may become true.

When counseling teams, I have generally found it more effective to focus on addressing whether an input variable can be controlled than trying to determine whether it will ever be controlled. This avoids the complications of having a team second-guess management.

GAINING INSIGHTS FROM THE PROCESS MAP

Figure 12.1 demonstrated the basic architecture and components of the process map. If we expand Figure 12.1 so that it represents a series of linked processes or even subprocesses, we obtain something similar to what is presented in Figure 12.2.

Though still simple in nature, Figure 12.2 depicts a representation of processes we are likely to encounter in any organization. For the sake of convenience, inputs and outputs in Figure 12.2 have been defined simply as letters. Since input variable classifications are not the focus of this section, they have been omitted.

Upon reviewing Figure 12.2, several immediate observations can be made, including:

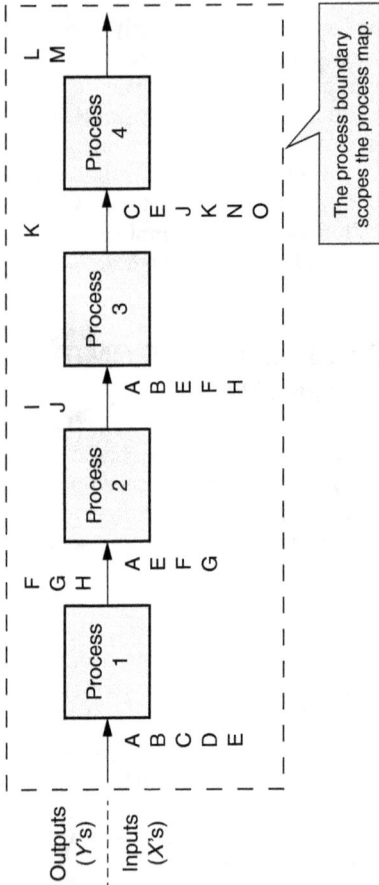

Figure 12.2   Analyzing inputs and outputs.

- Processes should have boundaries. Define them.

- Identical inputs may be required at different processes.

- Multiple process outputs may occur.

- Outputs of one process may become inputs of another process.

- Some outputs may or may not be linked to processes outside of the process boundaries.

We can now extend the thinking captured in the above bullets for each of the variables in Figure 12.2. These can be summarized easily in Table 12.3. Note the comments for each variable in Table 12.3. Notice that some of the comments are essentially a statement of fact, while others demand action, seek information, or require further investigation.

A well-developed process map serves as an effective communication tool and is a constant reminder of where a team should and should not focus its time and energy.

Table 12.3 Identifying potential issues with inputs and outputs.

Variable	Comment
A	Input to processes 1, 2, and 3. However, it is not an output from any of the processes within the scope of the project at hand. If it is output from another process outside the scope, the project team should ascertain its source to ensure its availability when required.
B	Input to processes 1 and 3. However, it is not an output from any of the processes within the scope of the project at hand. If it is output from another process outside the scope, the project team should ascertain its source to ensure its availability when required.
C	Input to processes 1 and 4. However, it is not an output from any of the processes within the scope of the project at hand. If it is output from another process outside the scope, the project team should ascertain its source to ensure its availability when required.
D	Input to process 1. However, it is not an output from any of the processes within the scope of the project at hand. If it is output from another process outside the scope, the project team should ascertain its source to ensure its availability when required.
E	Input to processes 1, 2, 3, and 4. However, it is not an output from any of the processes within the scope of the project at hand. If it is output from another process outside the scope, the project team should ascertain its source to ensure its availability when required.
F	Output from process 1 and input to both processes 2 and 3.

Continued

Table 12.3 *Continued.*

Variable	Comment
G	Output from process 1 and input to process 2.
H	Output from process 1 and input to process 3.
I	Output from process 2. This variable is no longer found in the remainder of the process map. It might be an extraneous output that is no longer needed, and process 2 was never changed to eliminate its production. Alternately, it could be used in another process beyond the scope of the project at hand. Either way, action is required.
J	Output from process 2 and input to process 4.
K	Output from process 3 and input to process 4.
L	Output from process 4. It may or may not be used in another process.
M	Output from process 4. It may or may not be used in another process.
N	Input to process 4. However, it is not an output from any of the processes within the scope of the project at hand. If it is output from another process outside the scope, the project team should ascertain its source to ensure its availability when required.
O	Input to process 4. However, it is not an output from any of the processes within the scope of the project at hand. If it is output from another process outside the scope, the project team should ascertain its source to ensure its availability when required.

Chapter 13

Data Collection

INTRODUCTION

At the outset of any data collection activity, operational definitions should be established for each data element that will be collected. This is to ensure the proper data elements will be collected and the opportunity for error will be minimized.

DEFINING AN OPERATIONAL DEFINITION

In the context of data collection and metric development, an operational definition is a clear, concise and unambiguous statement that provides a unified understanding of the data for all involved before the data are collected or the metric developed. It answers "Who collects the data, how are the data collected, what data are collected, where are the sources of the data, and when are the data collected?" Again, additional "who," "how," "what," "where," and "when" answers may be required.

When the data are used to construct a metric, the operational definition further defines the formula that will be used and each term used in the formula. As with data collection, the operational definition also delineates who provides the metrics and who is answerable to its results, how the metric is to be displayed or graphed, where the metric is displayed, and when it is available. Again, additional "who," "how," "what," "where," and "when" answers may be required.

Finally, the operational definition provides an interpretation of the metric, such as "up is good." While providing an interpretation of a metric seems trivial, this is not always the case.

For example, consider "employee attrition rate." How should this be interpreted? Is down good? Yes? How far down is "good"? Some authorities suggest that some amount of attrition is acceptable because it permits new thinking to enter the organization. A high level of attrition, however, can bleed the organization of institutional knowledge and memory.

In this example, the attrition rate metric may have boundaries in which:

- Above the upper boundaries—"down is good."

- Below the lower boundary—"up is good."

COLLECTING CUSTOMER DATA

This section will discuss the development of a customer-focused data collection strategy based on knowledge of the voice of the process and the voice of the customer. The discussion will include defining what it means to be customer focused and designing a customer data collection system.

Voice of the Process (VOP)

Typically, control charts are the mechanism by which we discover the voice of the process. In-control and stable processes provide us with estimates of natural process limits.

Natural process limits are known in the literature by several names: natural process variation, normal process variation, and natural tolerance. In all cases, these terms refer to the ±3σ limits (that is, 6σ spread) around a process average. Such limits include 99.73% of the process variation and are said to be the "*voice of the process.*" Walter Shewhart originally proposed the ±3σ limits as an economic trade-off between looking for special causes for points outside the control limits when no special causes existed and not looking for special causes when they did exist.

Voice of the Customer (VOC)

In contrast to natural process limits, specification limits are customer determined or derived from customer requirements and are used to define acceptable levels of process performance. Said to be the "*voice of the customer*," specification limits may be one-sided (that is, upper or lower) or two-sided. The difference between the upper and lower specification limits is known as the *tolerance*.

In some cases, customers provide explicit specifications for products or services. This is often the situation when the customer is the Department of Defense or any of the military branches. In others, customers express requirements in value terms, the components that influence the buy decision, such as price, product quality, innovation, service quality, company image, and reputation. In still other cases, customers may spotlight only their needs or wants, thus leaving it up to the organization to translate them into internal specifications. Tools such as quality function deployment, critical-to-quality analysis (also known as *customer requirements tree analysis*), and so on, often help with the last two situations.

Defining Customer-Focused

Organizations that recognize the importance of customers and focus on their wants and needs are commonly referred to as being customer focused or customer driven. In business literature, the terms customer driven and market driven are often used interchangeably. However, a customer-driven organization focuses more on the care and retention of existing customers, whereas the market-driven organization is more attuned to attracting new customers and serving their needs.

Designing a Customer-Based Data Collection System

Many organizations collect customer data but lack a systematic approach for doing so. Important data are often collected independently via many different departments but never shared, collated, reconciled, or analyzed to draw meaningful conclusions that drive action.

When designing a customer data collection system, the following key considerations should be at the forefront:

- *Collect data in a manner as objectively and consistently as possible.* Essentially, minimize measurement error. This concept should permeate all data collection mechanisms such as surveys, data collection check sheets, and so on. Consider using scripts for interviews to ensure that the same questions are asked the same way each time.

- *Collect data at the right level of granularity.* Consider what type of analysis will be done, what questions you are trying to answer, or the type of requirements you are trying to discover. If data are collected at too high a level, information is lost. Data collected at too low a level are meaningless. Fortunately, the data often can be summarized to an appropriate higher level for analysis. However, this creates additional work and consumes time.

- *Consider independent sources of data collection.* Multiple data collection mechanisms or listening posts should be used, as shown in Figure 13.1. This permits data regarding each customer category or segment to be validated. If the data can not be validated, this is generally a signal that improvement in the mechanisms (that is, revised questions) is required.

- *Use multiple media for collecting data.* Different customer categories or segments may favor different media for providing

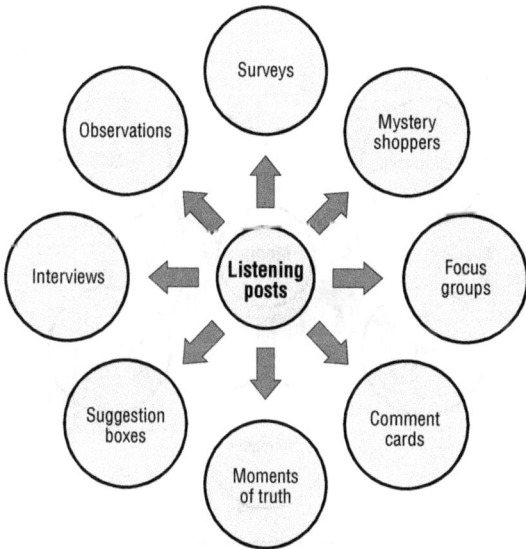

Figure 13.1 Examples of common listening posts.

feedback, as shown in Figure 13.2. The
key is to determine how to best align
listening posts with the most appropriate
media.

- *Make it easy for the customer.* Customers
  must have easy access to the organization
  to provide feedback. Similar to the multiple
  media bullet above, this bullet speaks

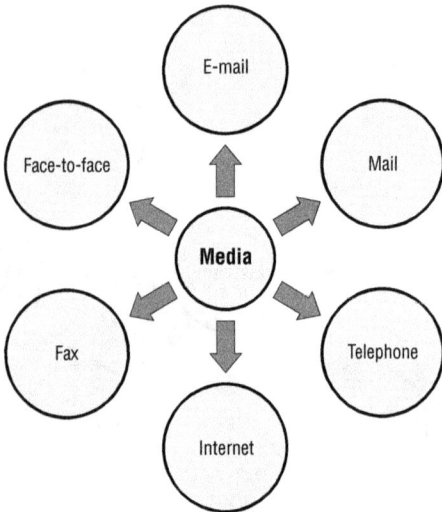

Figure 13.2 Examples of media used for listening posts.

to the ease with which the customer has access to the organization. For example, an organization uses an automated telephone system for collecting customer feedback. Customers may become frustrated with attempting to determine the correct routing path to take to provide the feedback. Consequently, they decide their input and time is not worth the effort. As a result, the organization loses a source of valuable information regarding the performance of its products and services. Similarly, surveys that are too long or inappropriately designed may also cause customers to abandon participation.

The above points focus mainly on collecting customer data from the organization's own internal sources of information. Such data should include both internal and external customers. Table 13.1

Table 13.1 Sources of customer data.

Internal	External
Complaints	Research (that is, magazines, newspapers, trade journals)
Claim resolutions	
Warranty and guarantee usage	Public information (that is, customers' and competitors' annual reports)
Service records	Advertising media (that is, television, radio, websites)
Customer-contact employees	Industry market research
Listening post data	Customers of competitors
Market research	Industry conferences and forums
Transaction data	
Defection data	

identifies these multiple sources of customer data. Notice that there are many internal sources of customer data. Actually, many organizations are data rich with regard to such information, but often fail to tap into it or even recognize it as an important source.

DATA INTEGRITY AND RELIABILITY

As quality or Lean Six Sigma professionals, we have been taught to address the issue of data accuracy and integrity from the statistical viewpoint. However, this section will address most of the elements of data integrity and reliability by discussing why data accuracy is important, the numerous causes of poor data accuracy, and various techniques for improvement and the placement of effective data collection points (Kubiak 2008).

Importance of Data Accuracy

Poor data accuracy undermines two key precepts in a quality-focused or Lean Six Sigma organization: fact-based and data-driven decision making. When faced with data that have been rendered useless because of data accuracy issues, decision makers must revert to relying on intuition.

This might provide the naysayer with reason and justification to state, "This Lean Six Sigma approach won't work here." Remember, Lean Six Sigma drives cultural changes in organizations, and accurate data are necessary to drive and sustain real improvement.

In addition, some organizations might be required to submit highly accurate data because of regulatory or reporting requirements, or because of contractual requirements. This is particularly true in industries such as aerospace, in which data, in addition to product, are a deliverable. In fact, most organizations have processes that deliver only data to other organizations.

Consider the example of payroll data regularly reported to the Internal Revenue Service. The consequences associated with poor data accuracy can be quite troublesome and have a strong negative impact on the individual depending on the data.

Minimizing Poor Data Accuracy

There are many causes of poor data accuracy. Table 13.2 includes a few of the most significant causes. Consider how these issues relate to your own experience and create your own list so you can minimize the impact of poor data accuracy in your organization.

Table 13.2   Common causes of poor data accuracy.

Issue	Comment
Multiple points of entry	When data entry is centralized to a single individual, entry errors are usually minimized over time due to the experience of the individual and, often, the ability of the individual to contact the originator of the data for clarifications. When data originators become responsible for their own data entry, data accuracy usually diminishes, particularly when frequency of data entry decreases or there is high turnover among the data originators.
Limited use of data validation	Such validation would include the presence or absence of data, data ranges, inconsistent data, and other logical checks. Surprisingly, computer technology has not completely eliminated this cause.
Batching input versus real-time input/ synchronization of system updates	This scenario usually results in instantaneous differences, but the differences are usually reconciled with time. Consider the example of an individual inquiring about a checking account balance through an ATM, a bank teller, or online. Because of different cut-off and system update times, the individual will see up to three different account balances.

Continued

Table 13.2 *Continued.*

Issue	Comment
Multiple means of data correction access to upstream systems	When multiple data correction access points are present, the opportunity to have synchronization issues increases. For example, suppose there are three upstream systems prior to the current system. Label them 4, 3, 2, and 1 for the current system. If corrections are made to 4, then 3, 2, and 1 are updated. However, if corrections are made to 2, only 2 and 1 are updated.
Unclear directions	This issue becomes increasingly important when multiple points of data entry exist.
Lack of training	This issue becomes increasingly important when multiple points of data entry exist.
Ambiguous terminology	Ask an operations manager for monthly data and you get monthly data according to the calendar month. Ask for the same data from an accountant and you get monthly data by fiscal month. Different months' lengths yield different data.
Manual versus automated means of entry	Automated data entry is usually preferred over manual data entry, particularly if you accept the premise that people are fallible.

Continued

Table 13.2 Continued.

Issue	Comment
Rounding	Improper rounding, for example, when rounding up is appropriate instead of rounding down. Also, the cumulative effects of rounding can generate errors. For example, computers store and calculate using numbers rounded to 16, 32, or even 64 digits. An individual using numbers rounded to four digits in the same calculations can arrive at a result that is substantially different than the results arrived at by a computer.
Order of calculations	Consider $\dfrac{36}{(1/9)}$. If you rewrite this equation as $36\left(\dfrac{9}{1}\right)$ and multiply through, you obtain 324. If you divide out 1/9 and round to four decimals, you obtain 0.1111. Dividing 36/0.1111 you obtain 324.0324. Different order, different answers.
Calculation errors	This is simply an arithmetic error or an error in the specification of the formula to be used. For example, I worked at two organizations that computed "associate turnover" incorrectly.
Inadequate measurement systems	This is an issue commonly known among Six Sigma professionals who have been taught to examine the adequacy of the measurement system first. Examples include Kappa studies and gage repeatability and reproducibility studies.

Continued

Table 13.2 *Continued.*

Issue	Comment
Units of measure not defined	This type of error occurs when the wrong use of measure is assumed or used, such as "foot" instead of "inch," "yard" instead "foot," or "box" instead of "each." The last example can significantly impact inventory levels.
Failing to use the proper system of measurement	On September 23, 1999, after a 286-day journey from Earth to Mars, NASA lost the $125 million Mars Climate Orbiter after the spacecraft entered orbit 100 kilometers closer than planned. NASA used the metric system, while its partner Lockheed Martin, the organization that designed the navigation system, used English units.
Field inconsistencies/ truncation among systems or data fields	Moving an n character field to an $n - 1$ character field or smaller can result in field truncation, which would impact all calculations that use it. For example, 100,789 is moved into a three-character numeric field. The result would be 789, not 100,789.

Continued

Table 13.2   *Continued.*

Issue	Comment
Similarities of characters	Several letters and numbers can easily be confused, particularly when handwriting is poor or when reading other than the top sheet of a no carbon required form. Common examples include:
	0 O (oh) 4 H 8 B
	1 I (el) 5 S 9 g (gee) or q
	2 Z and occasionally Q 6 G
	3 B 7 Z
	Examples of possible confusion between letters: c-e, C-O, E-F, m-n, M-N, O-Q, P-R, r-v, u-v, U-V, v-w, V-W, v-y
Copying errors	Duplicates of duplicates can result in image degradation and impact the readability of the copy. Also, copy truncation occurs when the image being copied is close to the edge of the paper. Try this with your credit card statement sometime. Ensure that the purchase amounts are adjacent to the edge of the copier glass.

Table 13.3 lists data collection techniques used to minimize the occurrence of poor data accuracy or its impact.

Remember, processes are measurable and generate data. Treat poor data accuracy as you would any other quality defect: use the plethora of tools available to you as a quality or Lean Six Sigma professional to ferret out root cause. Chart data defects as you would product defects. Make them available for all to see.

Common Data Collection Points

Five common data collection points are summarized in Table 13.4 and are as follows:

• *End of process.* Traditionally, organizations have collected data to measure process performance at the very end when everything is said and done, and it's usually too late to take corrective action.

• *In process.* Over time, organizations that are more progressive have realized that data should be collected upstream and during the process. Data collection at critical subprocesses allows corrective action to be taken earlier, when change is still an option. This move facilitates more-effective process management and allows the organizations to better gauge the overall health of their processes.

Table 13.3   Useful data collection techniques.

Technique	Comment
Walk the process first	Always walk the process first so everything that follows in this table has meaning and applicability.
Make the data collection process simple	Minimize the need to perform calculations and maximize the use of direct data collection. Direct data collection occurs when you take a reading from an instrument and log or enter that piece of data. In this case, a simple data transference has occurred.
Define collection points	Table 13.4 identifies typical and useful data collection points.
Use check sheets and checklists	These are simple but very effective tools and often overlooked. Perhaps they are too simple.
Bridge computer gaps	Some organizations still have processes requiring individuals to extract data from one computer report and enter it into a database residing on a different computer. In some cases, it is the same computer.

Continued

Table 13.3 *Continued.*

Technique	Comment
Minimize "other"	What is "other?" When collecting nominal or categorical data, it is often the last, and sometimes the largest, bar on a Pareto chart. A substantial amount of data classified as "other" usually indicates that data originators either don't understand clearly how to classify the data, or that categories are not mutually exclusive. The opportunity exists to classify data into multiple categories. As a result, data originators might disagree with one another, and the measurement system becomes inadequate.
Limit options	The more categories of data there are, the greater the possibility of a data collection error. This technique should be balanced against the previous "minimize other" technique.
Establish the rules	This technique is best illustrated by an example. Consider an organization that receives payments stating that any receipt received after 3:00 p.m. will be credited on the next business day.
Address timing and sequencing	This technique is important when a data collection point is inserted to collect data from multiple input streams. In this case, be particularly aware of the possibility of generating a bottleneck by waiting for a specific data input stream. Otherwise, it might be necessary to collect the data at another point downstream. If this is not possible, rebalancing the upstream processes might be required.

Continued

Table 13.3   *Continued.*

Technique	Comment
Time stamp data	This technique facilitates traceability of activities and events through time-based data and allows for more thorough root cause analysis.
Red tag a unit of product or service	Track a specified unit of product or service through the entire process to understand how it is used to generate data or how data are collected from it.
Chart data accuracy	Remember, data collection is part of the process and therefore subject to charting techniques such as control charts and run charts.

Table 13.4 Common data collection points (DCPs).

Point	Example
End of process	
In process	
Points of convergence	
Points of divergence	
Across functional boundaries	

• *Points of convergence*. This occurs when multiple streams of processes come together. Collecting data at this point makes sense because it helps organizations understand how each input stream affects the downstream subprocesses.

• *Points of divergence*. This is nothing more than a flip of the points of convergence. Keep in mind that a decision point is a specialized point of divergence with two distinct outputs.

Decision points involving "yes or no" decisions are useful data collection points, particularly when the "yes" path is the desired path and the "no" path is the undesired path.

Understanding the percentage of products or services taking the "no" path becomes an important piece of data necessary for driving process improvement.

• *Across functional boundaries*. These are nothing more than process hand-off points. However, responsibilities and accountability usually change across functional boundaries.

Collecting data at these focus points helps organizations understand how each functional department supports or detracts from process efficiency and effectiveness. It also helps minimize organizational finger-pointing with data and fact.

Accuracy Is Essential

Data accuracy is essential for managing and improving processes and achieving sustained results. Without it, decision making is, at best, uninformed.

Consequently, poor data accuracy can adversely impact an organization. Data defects, like product defects, should be charted and improved using the variety of tools at the disposal of the quality or Lean Six Sigma professional.

DATA COLLECTION STRATEGIES

Anderson-Cook and Borror (2013) identify seven data collection strategies, shown in Table 13.5. Notice that that each strategy is linked to one or more DMAIC phases. Furthermore, Anderson-Cook and Borror emphasize that the strategies are not mutually exclusive. Table 13.6 identifies the same strategies, but provides advantages and disadvantages of each.

Table 13.5 A summary of data collection strategies.

Collection strategy	DMAIC steps	Summary	Related strategies
1. Observational studies	D	• The relevance of the sample to the current study. Are there key differences in time, product, or process that may change patterns? • The quality of the data. How were the units selected? Have all potentially relevant inputs been gathered? • How the data can be used to advance the study without drawing conclusions that are not justifiable, given the lack of established causality.	2, 3, 4, 7
2. Monitoring techniques	A, C	• Selection of rational subgroups or individual observations to assess process stability • Action must be taken immediately (for example, when the monitoring technique is a control chart)	1, 3, 4, 5

Continued

Table 13.5 *Continued.*

Collection strategy	DMAIC steps	Summary	Related strategies
3. Process capability	A, C	Selection of rational subgroups or individual observations to assess process stability	1, 2, 5
4. Measurement assessment	M	Assessment measurement of system capability and stability	1, 2, 5, 6
5. Sampling	A, C	Intentional selection of subset of population	2, 3, 4, 7
6. Design of experiments	I	Active manipulation of factors to establish causality	4
7. Complementary data and information	D, A	Qualitative, brainstorming, expert opinion, and informal assessment	1, 5

Table 13.6   Advantages and disadvantages of the data collection strategies.

Collection strategy	Advantages	Disadvantages
1. Observational studies	• Conveniently available data • Usually a large amount of data from which to sample	• Assuming a pattern observed in a convenience sample will be present in the larger population • Assuming causation between a set of inputs and its effect on the responses • Missing a lurking variable that wasn't measured and is driving change in both the input values and the response • Sampling from a portion of the total population, which might give a biased view of the relationship.
2. Monitoring techniques	• If used properly, these techniques can be highly beneficial to the organization • Can signal the need for swift action to be taken	• It is easy to misuse or select the wrong monitoring technique (for example, when the monitoring technique is a control chart) • Can easily fall into disuse

Continued

Table 13.6 *Continued.*

Collection strategy	Advantages	Disadvantages
3. Process capability	• If used properly, this technique can be very effective in driving process improvement • Can provide estimates of short- and long-term variation	• It is easy to misuse or select the wrong process capability index • Much confusion exists around the use of this technique and associated indices such that the use of these indices might be a hard sel
4. Measurement assessment	• Provides insight into true process variation • Provides insight into the stability, bias, and linearity of the measurement system	• It is easy to misuse or select the wrong measurement assessment tool • Results may be difficult to interpret
5. Sampling	• Helps minimize costs • Can provide estimates of population parameters	• It is easy select the wrong sampling technique • Once implemented, it is easy for production personnel to shortcut the sampling plan

Continued

Table 13.6 *Continued.*

Collection strategy	Advantages	Disadvantages
6. Design of experiments	• Helps define the critical X's • Establishes the $Y = f(X)$ relationship	• It is easy to misuse or select the wrong experiment design • Can be expensive • May impact production
7. Complementary data and information	• Easy to teach and use • A little effort goes a long way	• Some techniques are subjective in nature

Chapter 14

Summaries of Key Tools and Techniques

FORCE-FIELD ANALYSIS

A well known, but little used, quality tool that is quite helpful in driving organizational change is the force-field analysis. In addition, it can be quite beneficial in helping complete the "Action plan" column of the stakeholder analysis form (Table 9.1).

Siebels (2004) defines *force-field analysis* as a "technique for analyzing the forces that aid or hinder an organization in reaching an objective." Such an objective could be cultural change.

Figure 14.1 illustrates the general format of a force-field analysis. The desired change is listed at the top in simple terms. Driving forces are listed on the left side of the vertical line. *Driving forces* are those forces that aid in achieving the objective. Similarly, restraining forces are placed on the right side of the line. *Restraining forces* are those forces that hinder or oppose the objective.

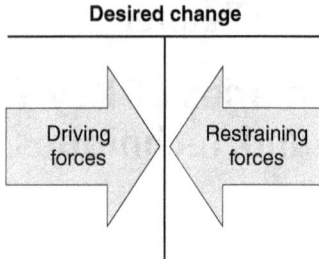

Figure 14.1 A force-field analysis diagram.

It is important to note that driving and restraining forces may be many-to-one in either direction, or even unopposed, as shown in Figure 14.2. When the unopposed force is a restraining force and is active, the reader may need to resort to the techniques discussed in Chapter 9. If these turn out to be ineffective, the sponsor may need to rethink the viability of the project.

Example 14.1

Figure 14.3 illustrates an example of losing weight provided by Tague (2005). As you review this example, notice that the driving and restraining forces are not necessarily or intended to be opposites. Also, notice that the relative strength of each force is not stated. This may be considered a drawback of the force-field analysis tool.

Driving forces   **Restraining forces**

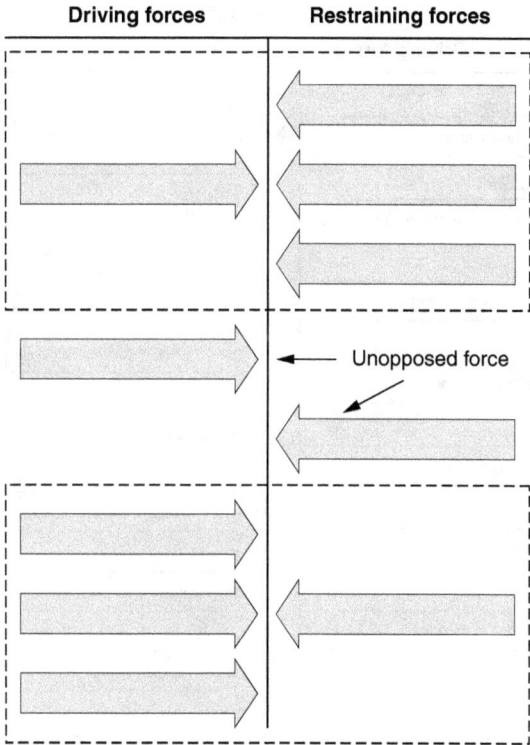

Figure 14.2 An example of many-to-one forces.

Losing weight

| Driving forces | Restraining forces |

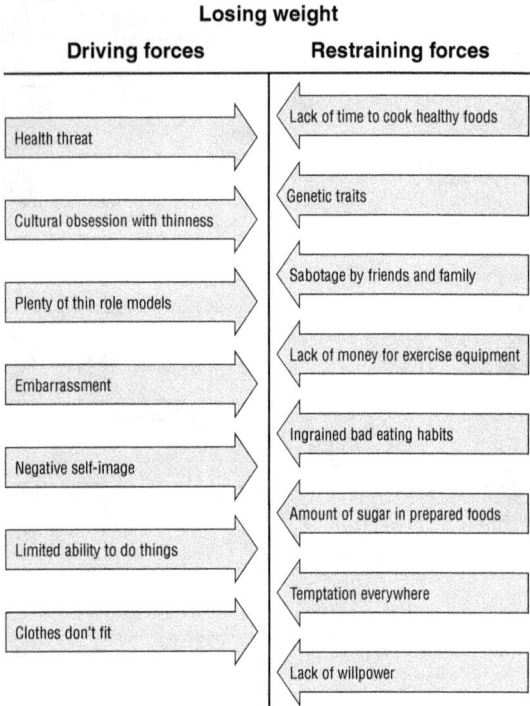

Figure 14.3 A force-field analysis diagram for Example 14.1.

Source: Adapted from Tague (2005).

Cultural maturity and readiness assessments can contain a wealth of information regarding both the driving and restraining forces. In addition, consider the use of brainstorming sessions. These can be conducted candidly and will likely surface additional and insightful information.

A force-field analysis can be done for change at any level, and is useful any time that there are two opposing sides to an issue that need to be considered. Identifying which forces can be changed and which can not will help reduce wasted energy. Focusing energy on removing or reducing forces that are restraining change, or ensuring that the driving forces are maintained or increased, will be more productive.

ACCURACY VERSUS PRECISION

Table 14.1 illustrates accuracy versus precision. By precision, we're really talking about variation. The top half of the table demonstrates more accuracy than precision, while the right half demonstrates more precision. Let's look at each quadrant of Table 14.1 as if it were a process being monitored by an $\bar{X} - R$ chart:

1. The mean of the process is accurate, but the precision is lacking and must be tightened.

Table 14.1   Accuracy versus precision.

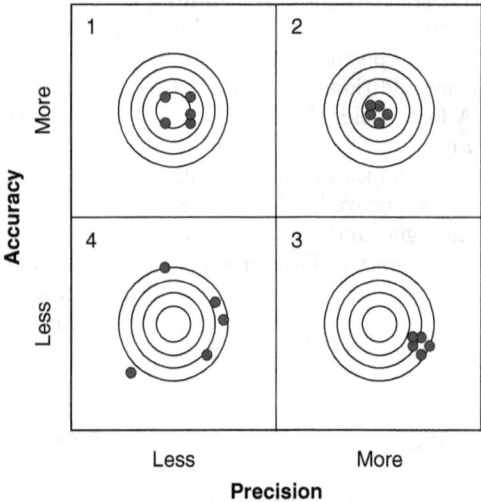

2. Both the mean and the precision of the process are very good.

3. The precision of the process is good, but the mean of the process must be adjusted.

4. Neither the mean nor the precision are very good. Both must be adjusted. Though some readers may argue that the mean

may average out to be on target, there is sufficient evidence to suggest otherwise.

CONTROL CHARTS

Control charts are statistically based graphical tools used to monitor the behavior of a process. Walter A. Shewhart developed them in the mid-1920s while working at Bell Laboratories. More than 80 years later, control charts continue to serve as the foundation for statistical quality control.

The graphical and statistical nature of control charts helps us:

- Quantify the variation of a process.

- Center a process.

- Monitor a process in relatively real time.

- Determine when or when not to take action on a process.

A wide variety of control charts have been developed over the years. However, we will focus on those that are the most popular in both manufacturing and transactional environments. These include the $\bar{X} - R$ chart, $\bar{X} - s$ chart, XmR chart, p-chart, np-chart, c-chart, and the u-chart.

The Structure of Control Charts

Constructing control charts is straightforward and, more often than not, aided by computer software designed specifically for this purpose. Minitab and JMP, among others, are commonly used.

Figure 14.4 illustrates the general form for a control chart. Its critical components are:

1. *X-axis*. This axis represents the time order of subgroups. Subgroups represent samples of data taken from a process. It is critical that the integrity of the time dimension be maintained when plotting control charts.

2. *Y-axis*. This axis represents the measured value of the quality characteristic under consideration when using variables charts. When attributes charts are used, this axis is used to quantify defectives or defects.

3. *Centerline*. The centerline represents the process average.

4. *Control limits*. Control limits typically appear at ±3σ from the process average.

5. *Zones*. The zones represent the distance between each standard deviation and

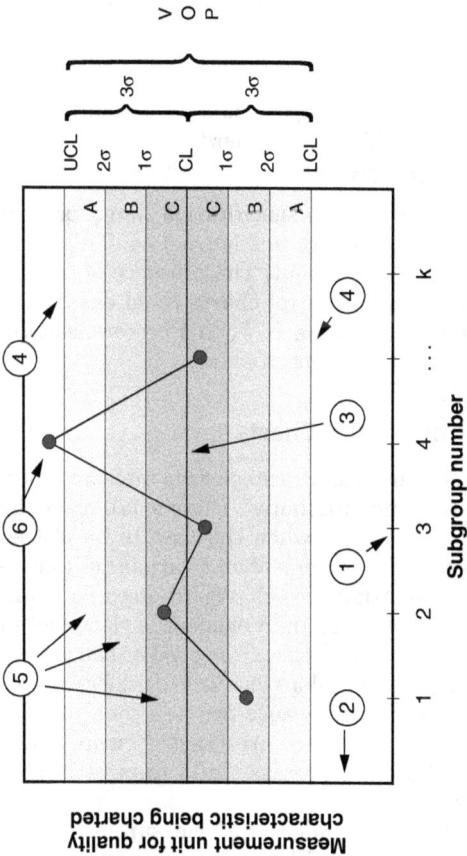

Figure 14.4 The structure of a typical control chart.

are useful when discussing specific out-of-control rules.

6. *Rational subgroups.* The variation within subgroups should be as small as possible so it's easier to detect subgroup-to-subgroup variation.

Notice there are no specification limits present on the control chart in Figure 14.4. This is by design, not by accident. The presence of specification limits on control charts could easily lead to inaction, particularly when a process is out of control but within specification.

Types of Control Charts

Control charts can be categorized into two types: variables and attributes. Charts fall into the variables category when the data to be plotted result from measurement on a variable or continuous scale. Attributes charts are used for count data in which each data element is classified in one of two categories, such as good or bad.

Generally, variables charts are preferred over attributes charts because the data contain more information, and they are typically more sensitive to detecting process shifts than attributes charts.

It should be noted that measurement data collected for use with a variables control chart can be categorized and applied to an attributes control chart. For example, consider five temperature readings. Each reading can be classified as being within specification (for example, good) or out of specification (for example, bad). Had each reading been classified as attribute data (for example, either within or out of specification) at the time of collection, and the actual measurement value not recorded, the attribute data could not be transformed for use on a variables control chart.

Variables Charts

The $\bar{X} - R$ chart is the flagship of variables control charts. It actually comprises two charts: the $\bar{X}$ chart, used to depict the process average, and the R chart, used to depict process variation using subgroup ranges.

The $\bar{X} - s$ chart is another variables control chart. With this chart, the sample standard deviation, s_i, for each subgroup is used to indicate process variation instead of the range. The standard deviation is a better measure of variation when the sample size is large (approximately 10 or larger).

The individuals moving range chart, XmR or ImR, is a useful chart when data are expensive to obtain or occur at a rate too slow to form rational subgroups. As the name implies, individual data points are plotted on one chart while the moving range (for example, absolute value of the difference between successive data points) is plotted on the moving range chart.

Figure 14.5 provides a flowchart for selecting the proper control chart based on the sample size.

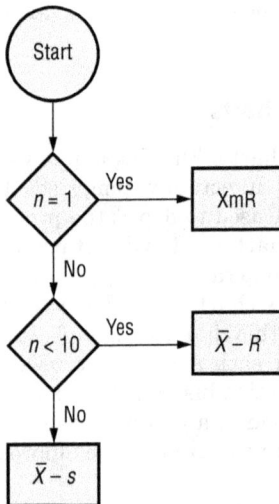

Figure 14.5   Selecting a variables control chart based on the subgroup size.

Attributes Charts

Attributes charts are used for count data in which each data element is classified in one of two categories, such as good or bad. The p-charts and np-charts are used to plot proportion defective and number of defectives, respectively. The c-charts and u-charts are used to plot counts of defects and defects per unit, respectively.

Both p-charts and np-charts are based on the binomial distribution. As such, it is assumed that the probability of a defective remains constant from subgroup to subgroup. Similarly, the c-charts and u-charts are based on the Poisson distribution. Therefore, the probability of the occurrence of a defect is assumed constant from subgroup to subgroup. When constructing attributes charts, it is important to keep these assumptions in mind to ensure statistical validity of the results.

Additionally, users of attributes charts should note that the np-chart and the c-chart require constant sample sizes. The p-chart and u-chart, however, permit variable sample sizes. As a result, the control limits for p-charts and u-charts will often appear ragged. These concepts are shown in Table 14.2.

Finally, users should understand the differing terminology that surrounds these charts. For example, a *defective* is also known as a

Table 14.2 Characteristics of attributes charts.

	Variable sample size	Constant sample size	
Defective (Nonconformance)	Proportion defective ***p***	Number of defectives ***np***	**Binomial**
Defect (Nonconformity)	Number of defects/unit ***u***	Number of defects ***c***	**Poisson**
	Ragged limits	**Fixed limits**	

nonconformance. Similarly, a *defect* is also known as a *nonconformity*.

Why the need for a different set of descriptors? The answer is likely rooted in the legal implications or consequences of using terms such as *defect* and *defective*. Though identical in the quality sense, terms like *nonconformity* and *nonconformance* tend to mitigate emotional reactions to less-than-perfect quality.

Control Limits

When first calculating control limits, it is prudent to collect as much data as practical. Many authors suggest at least 25 subgroups. The data are plotted and the charts reviewed for out-of-control conditions. If out-of-control conditions are found, root cause should be investigated and the associated data removed.

Consequently, a revised set of control limits should be computed. If it is not possible to determine root cause, the associated data should remain in the calculation for the limits. In this situation, it is likely the out-of-control condition was a statistical anomaly and really due to common cause variation. Common cause variation is addressed later in this section.

Because control limits are calculated based on data from the process, they represent the voice of the process (VOP). Typically, they are set at $\pm 3\sigma$. The upper control limit is designated as *UCL* while the lower control limit is designated as *LCL*. The difference between the UCL and the LCL constitutes a 6σ spread. This spread is known as the VOP and is a necessary value when determining process capability. See Figure 14.4.

Remember, ±3σ represents 99.73% of the data. The probability of a point falling outside the limits is only 0.27%. Shewhart felt these limits represented an economical trade-off of the consequences of looking for a special cause that doesn't exist and not looking for one when it does exist. This being said, users are free to set limits at values other than ±3σ. However, caution is recommended since the out-of-control rules given below no longer apply.

Tables 14.3 and 14.4 summarize the formulas for computing the central lines and control limits for variables control charts and attributes control charts, respectively. The constant values for $A_2, A_3, B_3, B_4, D_3, D_4$, and E_2 used in Table 14.3 can be found in Appendix 3.

A careful review of Tables 14.3 and 14.4 indicates the general form of the formula for setting control limits:

Process average ± (Constant) ×
(Measure of process variation)

It is important to note that calculating the LCL on the process variation chart could result in a negative value. If this should occur, the LCL is artificially set to zero because it is not possible to have negative ranges or standard deviations. Likewise, when an LCL computes to a negative value on attributes charts, the LCL is artificially

Table 14.3 Formulas for calculating the centerline and control limits for variables charts.

	Average chart		
Chart	**Centerline**	**LCL**	**UCL**
$\bar{X}$	$\bar{\bar{X}}$	$\bar{\bar{X}} - A_2\bar{R}$	$\bar{\bar{X}} + A_2\bar{R}$
R	$\bar{R}$	$D_3\bar{R}$	$D_4\bar{R}$
$\bar{X}$	$\bar{\bar{X}}$	$\bar{\bar{X}} - A_3\bar{s}$	$\bar{\bar{X}} + A_3\bar{s}$
s	$\bar{s}$	$B_3\bar{s}$	$B_4\bar{s}$

n = sample size of each subgroup

k = number of subgroups

$$\bar{X}_i = \frac{\sum_{j=1}^{n} x_{ij}}{n} = \text{Average of the } i\text{th subgroup, plot points}$$

$$\bar{\bar{X}} = \frac{\sum_{i=1}^{k} \bar{X}_i}{k} = \text{Centerline of the } \bar{X} \text{ chart}$$

R_i = Range of the ith subgroup, plot points

$$\bar{R} = \frac{\sum_{i=1}^{k} R_i}{k} = \text{Centerline of the } R \text{ chart}$$

s_i = Standard deviation of the ith subgroup

$$\bar{s} = \frac{\sum_{i=1}^{k} s_i}{k} = \text{Centerline of the } s \text{ chart}$$

Continued

Table 14.3   *Continued.*

	Individuals chart		
Chart	**Centerline**	**LCL**	**UCL**
I	$\bar{X}$	$\bar{X} - E_2\overline{mR}$	$\bar{X} + E_2\overline{mR}$
mR	$\overline{mR}$	$D_3\overline{mR}$	$D_4\overline{mR}$

k = Number of individual data measurements

x_i = Individual data measurements, plot points

$$\bar{X} = \frac{\sum\limits_{i=1}^{k} x_i}{k} = \text{Centerline of the } \bar{X} \text{ chart}$$

$mR_i = |x_i - x_{i-1}|$ for $i = 2, 3, \ldots, k$, plot points

$$\overline{mR} = \frac{\sum\limits_{i=2}^{k} mR_i}{k} = \text{Centerline for } mR \text{ chart (moving range)}$$

set to zero because, again, it is not possible to have a negative percentage defective or negative defect counts.

Using Control Charts

A control chart that has not triggered any out-of-control condition is considered stable and predictable, and operating in a state of statistical

Table 14.4 Formulas for calculating the centerline and control limits for attributes charts.

Chart	Centerline	LCL	UCL	Plot point
p	$\bar{p} = \dfrac{\sum\limits_{i=1}^{k} D_i}{\sum\limits_{i=1}^{k} n_i}$	$\mathrm{LCL}_p = \bar{p} - 3\sqrt{\dfrac{\bar{p}(1-\bar{p})}{n_i}}$	$\mathrm{UCL}_p = \bar{p} + 3\sqrt{\dfrac{\bar{p}(1-\bar{p})}{n_i}}$	$p_i = \dfrac{D_i}{n_i}$
np	$n\bar{p} = n\dfrac{\sum\limits_{i=1}^{k} D_i}{\sum\limits_{i=1}^{k} n_i}$	$\mathrm{LCL}_{np} = n\bar{p} - 3\sqrt{n\bar{p}(1-\bar{p})}$	$\mathrm{UCL}_{np} = n\bar{p} + 3\sqrt{n\bar{p}(1-\bar{p})}$	D_i
c	$\bar{c} = \dfrac{\sum\limits_{i=1}^{k} c_i}{k}$	$\mathrm{LCL}_c = \bar{c} - 3\sqrt{\bar{c}}$	$\mathrm{UCL}_c = \bar{c} + 3\sqrt{\bar{c}}$	c_i

Continued

Table 14.4   *Continued.*

Chart	Center line	LCL	UCL	Plot point
u	$\bar{u} = \dfrac{\sum\limits_{i=1}^{k} c_i}{\sum\limits_{i=1}^{k} n_i}$	$LCL_u = \bar{u} - 3\sqrt{\dfrac{\bar{u}}{n_i}}$	$UCL_u = \bar{u} + 3\sqrt{\dfrac{\bar{u}}{n_i}}$	$u_i = \dfrac{c_i}{n_i}$

Notes:

1. n = Sample size of each subgoup

2. k = Number of subgroups

3. D_i = Number of *defective* (nonconforming) units in the ith subgroup

4. c_i = Number of *defects* (nonconformities) in the ith subgroup

control. The variation depicted on the chart is due to common cause variation.

Points falling outside the limits or that meet any of the out-of-control rules outlined below are attributed to special cause variation. Such points, regardless of whether they constitute "good" or "bad" occurrences, should be investigated immediately while the cause-and-effect relationships and individual memories are fresh and access to documentation of process changes is readily available.

As the time between the out-of-control event and the beginning of the investigation increases, the likelihood of determining root causes diminishes greatly. Hence, the motto "time is of the essence" is most appropriate.

Finding root cause for out-of-control conditions may be frustrating and time-consuming, but the results are worthwhile. Ideally, root causes for good out-of-control conditions are incorporated into the process, while root causes for bad out-of-control conditions are removed.

Now a word of caution: Adjusting a process when it is not warranted by out-of-control conditions constitutes process tampering. This usually results in destabilizing a process, causing it to spiral out of control.

When variables charts are being used, the chart used to measure process variation (for example, R, s, and mR) should be reviewed first. Out-of-control conditions on this chart constitute changes of within-subgroup variation. Remember from rational subgrouping, we would like the variation of subgroups on this chart to be as small as possible because this measure of dispersion is used to compute the control limits on the corresponding process average chart (for example, $\bar{X}$ and X). The tighter the control limits on the process average chart, the easier it is to detect subgroup-to-subgroup variation.

Commonly used rules or tests have been devised to detect out-of-control conditions. The specific rules vary. Conceptually, however, they are similar. For example, the eight rules used by the software package Minitab are:

1. One point more than 3σ (beyond zone A) from the central line (on either side).

2. Nine points in a row on the same side of the central line.

3. Six points in a row, all increasing or all decreasing.

4. Fourteen points in a row, alternating up and down.

5. Two out of three points more than 2σ (zone A and beyond) from the central line (on the same side).

6. Four out of five points more than 1σ (zone B and beyond) from the central line (on the same side).

7. Fifteen points in a row within 1σ (within zone C) of the central line (on either side).

8. Eight points in a row more than 1σ (zone B and beyond) from the central line (on either side).

Generally, all of the above rules apply to variables control charts, while a subset of them applies to attributes control charts. If you are using Minitab to develop control charts, only those that apply to the specific control charts will be available for selection. Note that the fourth data point in Figure 14.4 represents an out-of-control point since it is in violation of rule 1.

The probabilities associated with the above out-of-control conditions occurring are both similar in value and relatively small. With the exception of points exceeding the control limits, most out-of-control conditions are subtle, and would likely go unnoticed without the aid of a computerized control chart.

Highly Effective Tool

Specific rules have been devised to determine when out-of-control conditions occur. These rules have been designed to permit us to detect early changes in a process, thus allowing us to take systematic action to discover the root cause of the variation or permit adjustment or other actions on the process before serious damage occurs.

Control charts operating in control are stable and predictable, and operating under the influence of common cause variation. Those operating with out-of-control conditions present are under the influence of special cause variation.

A control chart is relatively easy to develop and use, and it can be a highly effective statistical tool when selected properly and used correctly. Its selection and use alone, however, are not sufficient. When so indicated, control charts must be acted on in a timely manner so that root causes may be identified and removed from the process.

One last thing—when in doubt, avoid tampering with the process (Rooney et al. 2009).

TEAM MANAGEMENT

To be effective in obtaining results, the Black Belt must be effective in managing teams. To this end, the Black Belt must be familiar with team:

- Roles

- Preparation
- Motivation
- Dynamics
- Management

Team Roles

Teams seem to work best when members understand their assigned roles and the team dynamics. Some standard team roles together with typical duties include:

- *Champion.* The role of the champion was discussed in detail in Chapter 8. This individual does not generally attend regular team meetings.

- *Team leader.* This is commonly the Black Belt. The role of the Black Belt was discussed in detail in Chapter 8:

  1. Chairs team meetings and maintains team focus on the goal.

  2. Monitors progress toward the goal and communicates this to the organization.

  3. Manages administrative and record-keeping details.

4. Establishes and follows up on action assignments for individual team members.

- *Facilitator*

  1. Makes certain all members have an opportunity for input and that a full discussion of issues occurs. In some cases, the facilitator chairs the meeting.

  2. Helps the team leader keep the team on track.

  3. Summarizes progress using visual aids as appropriate.

  4. Provides methods for reaching decisions.

  5. Mitigates nonproductive behavior.

  6. Helps resolve conflicts.

- *Scribe/recorder.* This may be a temporary or rotating position:

  1. Maintains and publishes meeting minutes. May communicate action assignment reminders to individuals.

  2. May use visual aids to record discussion points so all can see and react.

- *Timekeeper*. This role is used when a timed agenda is used:

  1. Keeps track of the time when a timed agenda is used.

  2. Notifies the team leader or facilitator of the time remaining on each agenda item.

- *Coach*. The role of the coach was discussed in detail in Chapter 11:

  1. Works with the team leader and facilitator to move the team toward the objective.

  2. Helps provide resources for completion of team member assignments.

  3. Evaluates the team's progress.

- *Team member*

  1. Participates in team meetings.

  2. Communicates ideas and expertise.

  3. Listens openly to all ideas.

  4. Completes action assignments as scheduled.

It is not uncommon that some individuals find the role of "team member" to be troublesome.

More specifically, they find that they do not know how to behave in this role. As this information becomes known, it is imperative that the team leader take immediate action. Such action might include "team (member) training." This training would provide instruction regarding how to serve on a team, what is expected of a team member, and so on. Such training may be formal or informal depending on the extent of the need.

Team Preparation

Using teams to accomplish Lean Six Sigma projects has become commonplace. Team members and team dynamics bring resources to bear on a project that people working individually would not be able to produce. The best teams with the best intentions will perform suboptimally, however, without careful preparation. There seems to be no standard list of preparation steps, but the following items should be performed at a minimum:

- Set clear purposes and goals. These should be directly related to the project charter. Teams should not be asked to define their own purpose, although intermediate goals and timetables can be team generated. Teams often flounder when purposes are too vague. If team members have been selected to represent groups or supply technical knowledge, these roles should be announced at the initial or kickoff meeting. The initiation of the team should be in the form of a written document from an individual at the executive level of the organization. A separate document from each member's immediate supervisor is also very helpful in providing up-front recognition that the team member's participation is important for the success of the team.

- Team members need some basic team-building training. Without an understanding of how a team works and the individual behavior that advances team progress, the team will often get caught in personality concerns and turf wars.

- A schedule of team meetings should be published early, and team members should be asked to commit to attending all meetings. Additional subgroup meetings may be scheduled, but full team meetings should be held as originally scheduled.

- Teams can succeed only if management wants them to. If clear management support is not in place, the wisdom

of launching the team is in question.
Team members must know that they
have the authority to gather data, ask
difficult questions, and, in general,
think differently.

- All teams should have a sponsor who
  has a vested interest in the project. The
  sponsor reviews the team's progress,
  provides resources, and removes
  organizational roadblocks.

- The establishment of team norms is a
  helpful technique that is often initiated
  at the first team meeting. Team norms
  provide clear guidelines regarding what
  the team will and will not tolerate, and
  often define the consequences of violating
  the norms. Examples of team norms
  include:

  - Being on time. A consequence of being
    late may be to put a dollar into a
    kitty that will be used at the team's
    discretion at a later date.

  - Holding one conversation.

  - Demonstrating civility and courtesy
    to all members.

  - Accomplishing assigned tasks on
    time.

- Providing advance notice of not being able to attend a meeting.

- Participating in each meeting.

- Following a prepared agenda.

- Pulling the team back on track when it strays from the agenda topic.

There is an increase in the use of virtual teams— teams that hold team meetings via the internet or other electronic means. These teams require somewhat more attention to communication and documentation activities.

Team Motivation

In many enterprises, people feel like anonymous cogs in a giant machine. They become unwilling to invest much of themselves in a project. An important part of team leadership is generating and maintaining a high level of motivation toward project completion. The following techniques have proven successful:

- *Recognition.* Everyone appreciates recognition for his or her unique contributions. Some forms of recognition include:

  - Letters of appreciation sent to individuals and placed in personnel files

- – Public expressions of appreciation via meetings, newsletters, and so on

- – Tokens of appreciation such as trophies, gifts, or apparel

- *Rewards*. Monetary rewards are effective, especially when considerable personal time or sacrifice is involved. However, particular care must be taken with such rewards. Frequently, the individual giving the reward delivers the same dollar amount to each team member without consideration given to the amount of personal time, sacrifice, or results achieved by each team member. This is the easy route for the giver, but may generate hard feelings within the team, for they know the degree to which each team member participated. One way around this issue is to provide a set monetary limit for the entire team and allow the team to determine how to spend the money or divide it up.

- *Relationships within the team*. During team meetings, facilitators and other team members can enhance motivation by setting an example of civility and informality, relaxing outside roles, and exhibiting a willingness to learn from one another. Linking celebrations to the achievement of project milestones may enhance teamwork. Some authorities suggest that participation and motivation are improved by referring to team sessions as *workshops* rather than *meetings*.

Team Dynamics

Once teams have been formed, they must be built, because a true team is more than a collection of individuals. A team begins to take on a life of its own that is greater than the sum of its parts.

The building of a team begins with well-trained members who understand the roles and responsibilities of team members and how they may differ from their roles outside the team. For example, a first-line supervisor might find himself deferring to members of his group in a team session more than he would outside team meetings. Teams are built with understanding and trust in fellow team members; therefore, an early step in the building of a team should be informal introductions of each member.

Various devices are used for this process, such as "Tell your name, a non–job related interest, and what you hope the team accomplishes." This last request sometimes uncovers so-called hidden agendas. These are goals individuals have in addition to the team's stated goals and purposes. Whether or not such agendas are verbalized, team members must recognize their existence and importance. The most successful teams are able to suppress the hidden agendas in favor of team progress on the stated ones.

The best model for team leadership is the coach who strives to motivate all members to contribute their best. Productive teams occur when team coaches facilitate progress while recognizing and dealing with obstacles. Team leaders or team facilitators, or both, may perform the coaching function. Table 14.5 contains common team obstacles and associated solutions.

Time Management

Time is perhaps the most critical resource for any team. Team meeting time must be treated as the rare and valuable commodity that it is. It is up to the team leader and facilitator to make every minute count, although every team member has this responsibility as well. Some practices that have proven useful follow:

- Form an agenda committee with the function of generating the meeting agenda well in advance of the scheduled meeting. This group can take the responsibility for getting the resources called for by each agenda item. For smaller teams, the team leader often prepares the agenda.

Table 14.5 Team obstacles and solutions.

Obstacle	Solutions
• A person or group dominates the discussion	• Go around the team, asking each person for one comment or reaction.
	• Ask dominating people to summarize their positions or proposals and e-mail them to all team members.
	• If the dominating people tend to react negatively to the suggestions of others, ask them for their ideas first or adopt the "no judgments allowed" rule from the brainstorming technique.
	• Speak to the dominating people between team meetings, requesting their assistance and cooperation in making sure all voices are heard
• A person or group is reluctant to participate	• Be sure to welcome and express appreciation for every comment or contribution.
	• Form small subgroups that report to the full team.
	• Make action assignments for each person, with brief reports at the beginning of the next team meeting.
	• Speak to the reluctant people between team meetings, requesting their assistance and cooperation in making sure all voices are heard.

Continued

Table 14.5　*Continued.*

Obstacle	Solutions
• A tendency exists to accept opinions without data	• Emphasize the importance of basing decisions on facts from the first meeting onward.
	• Raise questions such as:
	– Are there data that can support that?
	– How do we know that?
	– How could we verify that?
	– Who could collect some data on that?
• Emphasis on consensus building has influenced a team to seek consensus too early (that is, before opposing views have had a fair hearing)	• Provide full opportunity for the expression of all views.
	• Support voices of dissent.
	• Ask individuals to play devil's advocate by opposing early consensus.
	• Be sure a preliminary written conclusion includes dissenting views.
• Team members begin to air old disputes	• Make sure ground rules state the need for a fresh start, keeping history in the past.
	• Have a printed agenda, possibly with scheduled times for agenda items. A facilitator may say, "It's 3:10; let's move on to item number two."
	• Assign team members the job of collecting data regarding the issue in dispute.

Continued

Table 14.5 *Continued.*

Obstacle	Solutions
• The team is floundering because it has lost sight of its goals and objectives	• Make sure goals and objectives are clear and well understood at the first team meeting. • Start team meetings by revisiting goals and objectives and offering a reminder as to where the team is in its journey and which objective is next. • Use graphics based on PERT and Gantt charts to help keep teams focused. • Bring in an outside voice, which often generates new ideas and approaches • Consider using a facilitator
• The team is rushing to meet its milestones or designated accomplishments without the benefit of a thorough study or analysis Note: When the above occurs, the team risks: • Sub-optimizing	• Encourage the team to study a far-ranging list of approaches and solutions. Use divergent thinking tools such as brainstorming and cause-and-effect diagrams to broaden the perspective. • When a change is studied, be sure it is clearly communicated to all who might be impacted. Actively seek input regarding possible unintended consequences. • If a root cause is "turned off," the problem should go away. Try toggling the root cause on and off and observe whether the problem is toggling also.

Continued

Table 14.5   *Continued.*

Obstacle	Solutions
• Unintended consequences (that is, solving one problem but creating others) • Missing root causes	
• The team becomes troubled over the issue of attribution (that is, who should get credit for an accomplishment)	• If an idea comes from outside the team, that should be acknowledged. • If an idea was advanced before the team was formed, that should be acknowledged. • Often, the source of ideas and concepts developed during team sessions is not known, but if one person or group was responsible, that should be acknowledged.
• The team digresses too far from its goals and objectives	• Assign an individual or group to follow through on a side topic. • Set a specific time as a deadline beyond which no effort will be spent on the topic.

- Use a Gantt chart timeline displaying milestones and dates, updating as needed.

- Publish meeting agendas with time limits on each item. Assign a timekeeper and stick to the schedule as closely as possible. Leave five to 10 minutes as a buffer at the end of the session.

- Team meetings should close with a review of activities individual team members will complete before the next meeting. Minutes of the meeting highlighting these activities should follow each meeting. This often works well in the form of an action register document. In some cases, the team leader will want to review progress on these assignments between meetings.

Appendix 1

Project Creation to Completion Process Definitions

Appendix 2 identifies the processes associated with a project from its creation to its completion. Six key important processes have been identified:

- *Project identification.* Project identification means the delineation of clear and specific improvement opportunities needed in the organization.

- *Project qualification.* Project qualification means the translation of an identified improvement opportunity into Lean Six Sigma charter format for recognizing its completeness and the ability to assess its potential as a viable project. (Note: Some organizations tend to combine the qualification and identification processes.)

- *Project selection.* Project selection means a project is now in suitable form and is ready to be judged against suitable criteria to determine its viability as a Lean Six Sigma project.

- *Project prioritization.* Project prioritization means that a selected project will be independently assessed against specific criteria in order to establish a rank score. Rank scores of other projects will be compared, and a final ranking of projects will be established. This final ranking will set the precedence and order for project assignment. (Note: Some organizations tend to combine the selection and prioritization processes.)

- *Project assignment.* Project assignment means that a prioritized project and a resource (that is, Black Belt or Black Belt candidate) are available to be matched. If the matching is successful, the resource is assigned to the project, and the project gets under way.

- *Project closure.* Project closure means that a project has been terminated for one of two reasons: successful completion or it is no longer viable.

It is important to note that some organizations will combine some of the above processes.

Appendix 2

Managing the Project Flow from Creation to Completion

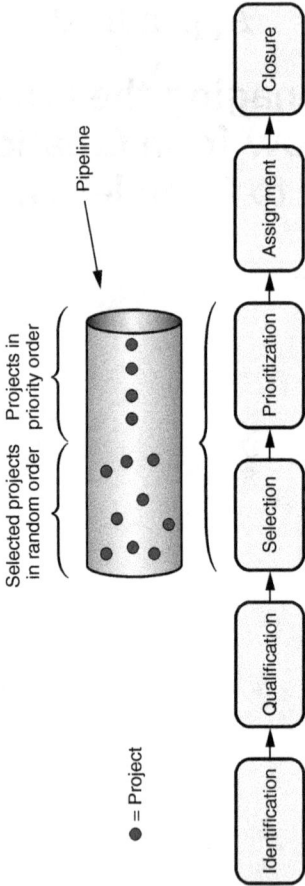

● = Project

Identification — Qualification — Selection — Prioritization — Assignment — Closure

Pipeline

Selected projects in random order — Projects in priority order

Appendix 3

Control Chart Constants

n	A_2	A_3	B_3	B_4	D_3	D_4	E_2	n
2	1.881	2.659	0.000	3.266	0.000	3.267	2.660	2
3	1.023	1.954	0.000	2.568	0.000	2.574	1.772	3
4	0.729	1.628	0.000	2.266	0.000	2.282	1.457	4
5	0.577	1.427	0.000	2.089	0.000	2.114	1.290	5
6	0.483	1.287	0.030	1.970	0.000	2.004	1.184	6
7	0.419	1.182	0.118	1.882	0.076	1.924	1.109	7
8	0.373	1.099	0.185	1.815	0.136	1.864	1.054	8
9	0.337	1.032	0.239	1.761	0.184	1.816	1.010	9
10	0.308	0.975	0.284	1.716	0.223	1.777	0.975	10
11	0.285	0.927	0.322	1.678	0.256	1.744	0.945	11
12	0.266	0.886	0.354	1.646	0.283	1.717	0.921	12
13	0.249	0.850	0.381	1.619	0.307	1.693	0.899	13
14	0.235	0.817	0.407	1.593	0.328	1.672	0.881	14
15	0.223	0.789	0.428	1.572	0.347	1.653	0.864	15
16	0.212	0.763	0.448	1.552	0.363	1.637	0.849	16
17	0.203	0.739	0.466	1.534	0.378	1.622	0.836	17
18	0.194	0.718	0.482	1.518	0.391	1.609	0.824	18
19	0.187	0.698	0.496	1.504	0.403	1.597	0.813	19

Continued

n	A_2	A_3	B_3	B_4	D_3	D_4	E_2	n
20	0.180	0.680	0.510	1.490	0.415	1.585	0.803	20
21	0.173	0.663	0.523	1.477	0.423	1.577	0.794	21
22	0.167	0.647	0.535	1.465	0.434	1.566	0.786	22
23	0.162	0.633	0.545	1.455	0.443	1.557	0.778	23
24	0.157	0.619	0.555	1.445	0.452	1.548	0.770	24
25	0.153	0.606	0.564	1.436	0.459	1.541	0.763	25

Appendix 4

Factors for Estimating σ_x

n	c_2	c_4	d_2	d_3	d_4	n
2	0.5642	0.7979	1.128	0.8525	0.954	2
3	0.7236	0.8862	1.693	0.8884	1.588	3
4	0.7979	0.9213	2.059	0.8798	1.978	4
5	0.8407	0.9400	2.326	0.8641	2.257	5
6	0.8686	0.9515	2.534	0.8480	2.472	6
7	0.8882	0.9594	2.704	0.8332	2.645	7
8	0.9027	0.9650	2.847	0.8198	2.791	8
9	0.9139	0.9693	2.970	0.8078	2.915	9
10	0.9227	0.9727	3.078	0.7971	3.024	10
11	0.9300	0.9754	3.173	0.7873	3.121	11
12	0.9359	0.9776	3.258	0.7785	3.207	12
13	0.9410	0.9794	3.336	0.7704	3.285	13
14	0.9453	0.9810	3.407	0.7630	3.356	14
15	0.9490	0.9823	3.472	0.7562	3.422	15
16	0.9523	0.9835	3.532	0.7499	3.482	16

Continued

n	c_2	c_4	d_2	d_3	d_4	n
17	0.9551	0.9845	3.588	0.7441	3.538	17
18	0.9576	0.9854	3.640	0.7386	3.591	18
19	0.9599	0.9862	3.689	0.7335	3.640	19
20	0.9619	0.9869	3.735	0.7287	3.686	20
21	0.9638	0.9876	3.778	0.7272	3.730	21
22	0.9655	0.9882	3.819	0.7199	3.771	22
23	0.9670	0.9887	3.858	0.7159	3.811	23
24	0.9684	0.9892	3.895	0.7121	3.847	24
25	0.9695	0.9896	3.931	0.7084	3.883	25

Note: σ_x may be estimated from k subgroups of size n:

$$\frac{\bar{\sigma}_{RMS}}{c_2} \qquad \frac{\bar{s}}{c_4}$$

$$\frac{\bar{R}}{d_2} \qquad \frac{\overline{mR}}{d_2}$$

$$\frac{\tilde{R}}{d_4}$$

Appendix 5

Standard Normal Distribution Table

$$\Pr(Z \geq z) = 1 - \Phi(Z \leq z) = \int_z^\infty \frac{1}{\sqrt{2\pi}} e^{-\mu^2/2} d\mu$$

Z	0.00	0.01	0.02	0.03	0.04	0.05	0.06	0.07	0.08	0.09	Z
0.0	0.5000	0.4960	0.4920	0.4880	0.4840	0.4801	0.4761	0.4721	0.4681	0.4641	0.0
0.1	0.4602	0.4562	0.4522	0.4483	0.4443	0.4404	0.4364	0.4325	0.4286	0.4247	0.1
0.2	0.4207	0.4168	0.4129	0.4090	0.4052	0.4013	0.3974	0.3936	0.3897	0.3859	0.2
0.3	0.3821	0.3783	0.3745	0.3707	0.3669	0.3632	0.3594	0.3557	0.3520	0.3483	0.3
0.4	0.3446	0.3409	0.3372	0.3336	0.3300	0.3264	0.3228	0.3192	0.3156	0.3121	0.4
0.5	0.3085	0.3050	0.3015	0.2981	0.2946	0.2912	0.2877	0.2843	0.2810	0.2776	0.5
0.6	0.2743	0.2709	0.2676	0.2643	0.2611	0.2578	0.2546	0.2514	0.2483	0.2451	0.6
0.7	0.2420	0.2389	0.2358	0.2327	0.2296	0.2266	0.2236	0.2206	0.2177	0.2148	0.7
0.8	0.2119	0.2090	0.2061	0.2033	0.2005	0.1977	0.1949	0.1922	0.1894	0.1867	0.8
0.9	0.1841	0.1814	0.1788	0.1762	0.1736	0.1711	0.1685	0.1660	0.1635	0.1611	0.9

Continued

Z	0.00	0.01	0.02	0.03	0.04	0.05	0.06	0.07	0.08	0.09	Z
1.0	0.1587	0.1562	0.1539	0.1515	0.1492	0.1469	0.1446	0.1423	0.1401	0.1379	1.0
1.1	0.1357	0.1335	0.1314	0.1292	0.1271	0.1251	0.1230	0.1210	0.1190	0.1170	1.1
1.2	0.1151	0.1131	0.1112	0.1093	0.1075	0.1056	0.1038	0.1020	0.1003	0.0985	1.2
1.3	0.0968	0.0951	0.0934	0.0918	0.0901	0.0885	0.0869	0.0853	0.0838	0.0823	1.3
1.4	0.0808	0.0793	0.0778	0.0764	0.0749	0.0735	0.0721	0.0708	0.0694	0.0681	1.4
1.5	0.0668	0.0655	0.0643	0.0630	0.0618	0.0606	0.0594	0.0582	0.0571	0.0559	1.5
1.6	0.0548	0.0537	0.0526	0.0516	0.0505	0.0495	0.0485	0.0475	0.0465	0.0455	1.6
1.7	0.0446	0.0436	0.0427	0.0418	0.0409	0.0401	0.0392	0.0384	0.0375	0.0367	1.7
1.8	0.0359	0.0351	0.0344	0.0336	0.0329	0.0322	0.0314	0.0307	0.0301	0.0294	1.8
1.9	0.0287	0.0281	0.0274	0.0268	0.0262	0.0256	0.0250	0.0244	0.0239	0.0233	1.9
2.0	0.0228	0.0222	0.0217	0.0212	0.0207	0.0202	0.0197	0.0192	0.0188	0.0183	2.0
2.1	0.0179	0.0174	0.0170	0.0166	0.0162	0.0158	0.0154	0.0150	0.0146	0.0143	2.1
2.2	0.0139	0.0136	0.0132	0.0129	0.0125	0.0122	0.0119	0.0116	0.0113	0.0110	2.2
2.3	0.0107	0.0104	0.0102	0.0099	0.0096	0.0094	0.0091	0.0089	0.0087	0.0084	2.3
2.4	0.0082	0.0080	0.0078	0.0075	0.0073	0.0071	0.0069	0.0068	0.0066	0.0064	2.4

Continued

Z	0.00	0.01	0.02	0.03	0.04	0.05	0.06	0.07	0.08	0.09	Z
2.5	0.0062	0.0060	0.0059	0.0057	0.0055	0.0054	0.0052	0.0051	0.0049	0.0048	2.5
2.6	0.0047	0.0045	0.0044	0.0043	0.0041	0.0040	0.0039	0.0038	0.0037	0.0036	2.6
2.7	0.0035	0.0034	0.0033	0.0032	0.0031	0.0030	0.0029	0.0028	0.0027	0.0026	2.7
2.8	0.0026	0.0025	0.0024	0.0023	0.0023	0.0022	0.0021	0.0021	0.0020	0.0019	2.8
2.9	0.0019	0.0018	0.0018	0.0017	0.0016	0.0016	0.0015	0.0015	0.0014	0.0014	2.9
3.0	0.0013	0.0013	0.0013	0.0012	0.0012	0.0011	0.0011	0.0011	0.0010	0.0010	3.0
3.1	0.00097	0.00094	0.00090	0.00087	0.00084	0.00082	0.00079	0.00076	0.00074	0.00071	3.1
3.2	0.00069	0.00066	0.00064	0.00062	0.00060	0.00058	0.00056	0.00054	0.00052	0.00050	3.2
3.3	0.00048	0.00047	0.00045	0.00043	0.00042	0.00040	0.00039	0.00038	0.00036	0.00035	3.3
3.4	0.00034	0.00032	0.00031	0.00030	0.00029	0.00028	0.00027	0.00026	0.00025	0.00024	3.4
3.5	0.00023	0.00022	0.00022	0.00021	0.00020	0.00019	0.00019	0.00018	0.00017	0.00017	3.5
3.6	0.00016	0.00015	0.00015	0.00014	0.00014	0.00013	0.00013	0.00012	0.00012	0.00011	3.6
3.7	0.00011	0.00010	0.00010	0.00010	0.00009	0.00009	0.00008	0.00008	0.00008	0.00008	3.7
3.8	0.00007	0.00007	0.00007	0.00006	0.00006	0.00006	0.00006	0.00005	0.00005	0.00005	3.8
3.9	0.00005	0.00005	0.00004	0.00004	0.00004	0.00004	0.00004	0.00004	0.00003	0.00003	3.9

Continued

Z	0.00	0.01	0.02	0.03	0.04	0.05	0.06	0.07	0.08	0.09	Z
4.0	0.00003	0.00003	0.00003	0.00003	0.00003	0.00003	0.00002	0.00002	0.00002	0.00002	4.0
4.1	0.00002	0.00002	0.00002	0.00002	0.00002	0.00002	0.00002	0.00002	0.00001	0.00001	4.1
4.2	0.00001	0.00001	0.00001	0.00001	0.00001	0.00001	0.00001	0.00001	0.00001	0.00001	4.2
4.3	0.00001	0.00001	0.00001	0.00001	0.00001	0.00001	0.00001	0.00001	0.00001	0.00001	4.3
4.4	0.000005	0.000005	0.000005	0.000005	0.000004	0.000004	0.000004	0.000004	0.000004	0.000004	4.4
4.5	0.000003	0.000003	0.000003	0.000003	0.000003	0.000003	0.000003	0.000002	0.000002	0.000002	4.5
4.6	0.000002	0.000002	0.000002	0.000002	0.000002	0.000002	0.000002	0.000002	0.000001	0.000001	4.6
4.7	0.000001	0.000001	0.000001	0.000001	0.000001	0.000001	0.000001	0.000001	0.000001	0.000001	4.7
4.8	0.000001	0.000001	0.000001	0.000001	0.000001	0.000001	0.000001	0.000001	0.000001	0.000001	4.8
4.9	0.000000	0.000000	0.000000	0.000000	0.000000	0.000000	0.000000	0.000000	0.000000	0.000000	4.9
5.0	0.000000	0.000000	0.000000	0.000000	0.000000	0.000000	0.000000	0.000000	0.000000	0.000000	5.0
5.1	0.000000	0.000000	0.000000	0.000000	0.000000	0.000000	0.000000	0.000000	0.000000	0.000000	5.1
5.2	0.000000	0.000000	0.000000	0.000000	0.000000	0.000000	0.000000	0.000000	0.000000	0.000000	5.2
5.3	0.000000	0.000000	0.000000	0.000000	0.000000	0.000000	0.000000	0.000000	0.000000	0.000000	5.3
5.4	0.000000	0.000000	0.000000	0.000000	0.000000	0.000000	0.000000	0.000000	0.000000	0.000000	5.4

Continued

Z	0.00	0.01	0.02	0.03	0.04	0.05	0.06	0.07	0.08	0.09	Z
5.5	0.000000	0.000000	0.000000	0.000000	0.000000	0.000000	0.000000	0.000000	0.000000	0.000000	5.5
5.6	0.000000	0.000000	0.000000	0.000000	0.000000	0.000000	0.000000	0.000000	0.000000	0.000000	5.6
5.7	0.000000	0.000000	0.000000	0.000000	0.000000	0.000000	0.000000	0.000000	0.000000	0.000000	5.7
5.8	0.000000	0.000000	0.000000	0.000000	0.000000	0.000000	0.000000	0.000000	0.000000	0.000000	5.8
5.9	0.000000	0.000000	0.000000	0.000000	0.000000	0.000000	0.000000	0.000000	0.000000	0.000000	5.9
6.0	0.000000	0.000000	0.000000	0.000000	0.000000	0.000000	0.000000	0.000000	0.000000	0.000000	6.0

Appendix 6

Cumulative Standard Normal Distribution Table

$$\Pr(Z \le z) = \Phi(Z \le z) = \int_{-\infty}^{z} \frac{1}{\sqrt{2\pi}} e^{-\mu^2/2} d\mu$$

Z	0.00	0.01	0.02	0.03	0.04	0.05	0.06	0.07	0.08	0.09	Z
0.0	0.5000	0.5040	0.5080	0.5120	0.5160	0.5199	0.5239	0.5279	0.5319	0.5359	0.0
0.1	0.5398	0.5438	0.5478	0.5517	0.5557	0.5596	0.5636	0.5675	0.5714	0.5753	0.1
0.2	0.5793	0.5832	0.5871	0.5910	0.5948	0.5987	0.6026	0.6064	0.6103	0.6141	0.2
0.3	0.6179	0.6217	0.6255	0.6293	0.6331	0.6368	0.6406	0.6443	0.6480	0.6517	0.3
0.4	0.6554	0.6591	0.6628	0.6664	0.6700	0.6736	0.6772	0.6808	0.6844	0.6879	0.4
0.5	0.6915	0.6950	0.6985	0.7019	0.7054	0.7088	0.7123	0.7157	0.7190	0.7224	0.5
0.6	0.7257	0.7291	0.7324	0.7357	0.7389	0.7422	0.7454	0.7486	0.7517	0.7549	0.6
0.7	0.7580	0.7611	0.7642	0.7673	0.7704	0.7734	0.7764	0.7794	0.7823	0.7852	0.7
0.8	0.7881	0.7910	0.7939	0.7967	0.7995	0.8023	0.8051	0.8078	0.8106	0.8133	0.8
0.9	0.8159	0.8186	0.8212	0.8238	0.8264	0.8289	0.8315	0.8340	0.8365	0.8389	0.9

Continued

Z	0.00	0.01	0.02	0.03	0.04	0.05	0.06	0.07	0.08	0.09	Z
1.0	0.8413	0.8438	0.8461	0.8485	0.8508	0.8531	0.8554	0.8577	0.8599	0.8621	1.0
1.1	0.8643	0.8665	0.8686	0.8708	0.8729	0.8749	0.8770	0.8790	0.8810	0.8830	1.1
1.2	0.8849	0.8869	0.8888	0.8907	0.8925	0.8944	0.8962	0.8980	0.8997	0.9015	1.2
1.3	0.9032	0.9049	0.9066	0.9082	0.9099	0.9115	0.9131	0.9147	0.9162	0.9177	1.3
1.4	0.9192	0.9207	0.9222	0.9236	0.9251	0.9265	0.9279	0.9292	0.9306	0.9319	1.4
1.5	0.9332	0.9345	0.9357	0.9370	0.9382	0.9394	0.9406	0.9418	0.9429	0.9441	1.5
1.6	0.9452	0.9463	0.9474	0.9484	0.9495	0.9505	0.9515	0.9525	0.9535	0.9545	1.6
1.7	0.9554	0.9564	0.9573	0.9582	0.9591	0.9599	0.9608	0.9616	0.9625	0.9633	1.7
1.8	0.9641	0.9649	0.9656	0.9664	0.9671	0.9678	0.9686	0.9693	0.9699	0.9706	1.8
1.9	0.9713	0.9719	0.9726	0.9732	0.9738	0.9744	0.9750	0.9756	0.9761	0.9767	1.9
2.0	0.9772	0.9778	0.9783	0.9788	0.9793	0.9798	0.9803	0.9808	0.9812	0.9817	2.0
2.1	0.9821	0.9826	0.9830	0.9834	0.9838	0.9842	0.9846	0.9850	0.9854	0.9857	2.1
2.2	0.9861	0.9864	0.9868	0.9871	0.9875	0.9878	0.9881	0.9884	0.9887	0.9890	2.2
2.3	0.9893	0.9896	0.9898	0.9901	0.9904	0.9906	0.9909	0.9911	0.9913	0.9916	2.3
2.4	0.9918	0.9920	0.9922	0.9925	0.9927	0.9929	0.9931	0.9932	0.9934	0.9936	2.4

Continued

Z	0.00	0.01	0.02	0.03	0.04	0.05	0.06	0.07	0.08	0.09	Z
2.5	0.9938	0.9940	0.9941	0.9943	0.9945	0.9946	0.9948	0.9949	0.9951	0.9952	2.5
2.6	0.9953	0.9955	0.9956	0.9957	0.9959	0.9960	0.9961	0.9962	0.9963	0.9964	2.6
2.7	0.9965	0.9966	0.9967	0.9968	0.9969	0.9970	0.9971	0.9972	0.9973	0.9974	2.7
2.8	0.9974	0.9975	0.9976	0.9977	0.9977	0.9978	0.9979	0.9979	0.9980	0.9981	2.8
2.9	0.9981	0.9982	0.9982	0.9983	0.9984	0.9984	0.9985	0.9985	0.9986	0.9986	2.9
3.0	0.9987	0.9987	0.9987	0.9988	0.9988	0.9989	0.9989	0.9989	0.9990	0.9990	3.0
3.1	0.99903	0.99906	0.99910	0.99913	0.99916	0.99918	0.99921	0.99924	0.99926	0.99929	3.1
3.2	0.99931	0.99934	0.99936	0.99938	0.99940	0.99942	0.99944	0.99946	0.99948	0.99950	3.2
3.3	0.99952	0.99953	0.99955	0.99957	0.99958	0.99960	0.99961	0.99962	0.99964	0.99965	3.3
3.4	0.99966	0.99968	0.99969	0.99970	0.99971	0.99972	0.99973	0.99974	0.99975	0.99976	3.4
3.5	0.99977	0.99978	0.99978	0.99979	0.99980	0.99981	0.99981	0.99982	0.99983	0.99983	3.5
3.6	0.99984	0.99985	0.99985	0.99986	0.99986	0.99987	0.99987	0.99988	0.99988	0.99989	3.6
3.7	0.99989	0.99990	0.99990	0.99990	0.99991	0.99991	0.99992	0.99992	0.99992	0.99992	3.7
3.8	0.99993	0.99993	0.99993	0.99994	0.99994	0.99994	0.99994	0.99995	0.99995	0.99995	3.8
3.9	0.99995	0.99995	0.99996	0.99996	0.99996	0.99996	0.99996	0.99996	0.99997	0.99997	3.9

Continued

Z	0.00	0.01	0.02	0.03	0.04	0.05	0.06	0.07	0.08	0.09	Z
4.0	0.99997	0.99997	0.99997	0.99997	0.99997	0.99997	0.99998	0.99998	0.99998	0.99998	4.0
4.1	0.99998	0.99998	0.99998	0.99998	0.99998	0.99998	0.99998	0.99998	0.99999	0.99999	4.1
4.2	0.99999	0.99999	0.99999	0.99999	0.99999	0.99999	0.99999	0.99999	0.99999	0.99999	4.2
4.3	0.99999	0.99999	0.99999	0.99999	0.99999	0.99999	0.99999	0.99999	0.99999	0.99999	4.3
4.4	0.999995	0.999995	0.999995	0.999995	0.999996	0.999996	0.999996	0.999996	0.999996	0.999996	4.4
4.5	0.999997	0.999997	0.999997	0.999997	0.999997	0.999997	0.999997	0.999998	0.999998	0.999998	4.5
4.6	0.999998	0.999998	0.999998	0.999998	0.999998	0.999998	0.999998	0.999998	0.999999	0.999999	4.6
4.7	0.999999	0.999999	0.999999	0.999999	0.999999	0.999999	0.999999	0.999999	0.999999	0.999999	4.7
4.8	0.999999	0.999999	0.999999	0.999999	0.999999	0.999999	0.999999	0.999999	0.999999	0.999999	4.8
4.9	1.000000	1.000000	1.000000	1.000000	1.000000	1.000000	1.000000	1.000000	1.000000	1.000000	4.9
5.0	1.000000	1.000000	1.000000	1.000000	1.000000	1.000000	1.000000	1.000000	1.000000	1.000000	5.0
5.1	1.000000	1.000000	1.000000	1.000000	1.000000	1.000000	1.000000	1.000000	1.000000	1.000000	5.1
5.2	1.000000	1.000000	1.000000	1.000000	1.000000	1.000000	1.000000	1.000000	1.000000	1.000000	5.2
5.3	1.000000	1.000000	1.000000	1.000000	1.000000	1.000000	1.000000	1.000000	1.000000	1.000000	5.3
5.4	1.000000	1.000000	1.000000	1.000000	1.000000	1.000000	1.000000	1.000000	1.000000	1.000000	5.4

Continued

Z	0.00	0.01	0.02	0.03	0.04	0.05	0.06	0.07	0.08	0.09	Z
5.5	1.000000	1.000000	1.000000	1.000000	1.000000	1.000000	1.000000	1.000000	1.000000	1.000000	5.5
5.6	1.000000	1.000000	1.000000	1.000000	1.000000	1.000000	1.000000	1.000000	1.000000	1.000000	5.6
5.7	1.000000	1.000000	1.000000	1.000000	1.000000	1.000000	1.000000	1.000000	1.000000	1.000000	5.7
5.8	1.000000	1.000000	1.000000	1.000000	1.000000	1.000000	1.000000	1.000000	1.000000	1.000000	5.8
5.9	1.000000	1.000000	1.000000	1.000000	1.000000	1.000000	1.000000	1.000000	1.000000	1.000000	5.9
6.0	1.000000	1.000000	1.000000	1.000000	1.000000	1.000000	1.000000	1.000000	1.000000	1.000000	6.0

Appendix 7

t Distribution Table

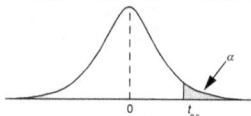

| ν | α | | | | | ν |
	0.10	0.05	0.025	0.01	0.005	
1	3.078	6.314	12.706	31.821	63.657	1
2	1.886	2.920	4.303	6.965	9.925	2
3	1.638	2.353	3.182	4.541	5.841	3
4	1.533	2.132	2.776	3.747	4.604	4
5	1.476	2.015	2.571	3.365	4.032	5
6	1.440	1.943	2.447	3.143	3.707	6
7	1.415	1.895	2.365	2.998	3.499	7
8	1.397	1.860	2.306	2.896	3.355	8
9	1.383	1.833	2.262	2.821	3.250	9
10	1.372	1.812	2.228	2.764	3.169	10
11	1.363	1.796	2.201	2.718	3.106	11
12	1.356	1.782	2.179	2.681	3.055	12
13	1.350	1.771	2.160	2.650	3.012	13

Continued

	α					
ν	0.10	0.05	0.025	0.01	0.005	ν
14	1.345	1.761	2.145	2.624	2.977	14
15	1.341	1.753	2.131	2.602	2.947	15
16	1.337	1.746	2.120	2.583	2.921	16
17	1.333	1.740	2.110	2.567	2.898	17
18	1.330	1.734	2.101	2.552	2.878	18
19	1.328	1.729	2.093	2.539	2.861	19
20	1.325	1.725	2.086	2.528	2.845	20
21	1.323	1.721	2.080	2.518	2.831	21
22	1.321	1.717	2.074	2.508	2.819	22
23	1.319	1.714	2.069	2.500	2.807	23
24	1.318	1.711	2.064	2.492	2.797	24
25	1.316	1.708	2.060	2.485	2.787	25
26	1.315	1.706	2.056	2.479	2.779	26
27	1.314	1.703	2.052	2.473	2.771	27
28	1.313	1.701	2.048	2.467	2.763	28
29	1.311	1.699	2.045	2.462	2.756	29
30	1.310	1.697	2.042	2.457	2.750	30
40	1.303	1.684	2.021	2.423	2.704	40
50	1.299	1.676	2.009	2.403	2.678	50
60	1.296	1.671	2.000	2.390	2.660	60
70	1.294	1.667	1.994	2.381	2.648	70
80	1.292	1.664	1.990	2.374	2.639	80
90	1.291	1.662	1.987	2.368	2.632	90
100	1.290	1.660	1.984	2.364	2.626	100
120	1.289	1.658	1.980	2.358	2.617	120
∞	1.282	1.645	1.960	2.326	2.576	∞

Appendix 8

Chi-Square
Distribution Table

ν	0.995	0.990	0.975	0.950	0.900	0.100	0.050	0.025	0.010	0.005	ν
1	0.00004	0.00016	0.00098	0.00393	0.01579	2.7055	3.8415	5.0239	6.6349	7.8794	1
2	0.010	0.020	0.051	0.103	0.211	4.605	5.991	7.378	9.210	10.597	2
3	0.072	0.115	0.216	0.352	0.584	6.251	7.815	9.348	11.345	12.838	3
4	0.207	0.297	0.484	0.711	1.064	7.779	9.488	11.143	13.277	14.860	4
5	0.412	0.554	0.831	1.145	1.610	9.236	11.070	12.833	15.086	16.750	5
6	0.676	0.872	1.237	1.635	2.204	10.645	12.592	14.449	16.812	18.548	6
7	0.989	1.239	1.690	2.167	2.833	12.017	14.067	16.013	18.475	20.278	7
8	1.344	1.646	2.180	2.733	3.490	13.362	15.507	17.535	20.090	21.955	8
9	1.735	2.088	2.700	3.325	4.168	14.684	16.919	19.023	21.666	23.589	9
10	2.156	2.558	3.247	3.940	4.865	15.987	18.307	20.483	23.209	25.188	10

The figure shows the chi-square distribution with the upper tail area α to the right of $\chi^2_{\alpha,\nu}$.

Continued

α

ν	0.995	0.990	0.975	0.950	0.900	0.100	0.050	0.025	0.010	0.005	ν
11	2.603	3.053	3.816	4.575	5.578	17.275	19.675	21.920	24.725	26.757	11
12	3.074	3.571	4.404	5.226	6.304	18.549	21.026	23.337	26.217	28.300	12
13	3.565	4.107	5.009	5.892	7.042	19.812	22.362	24.736	27.688	29.819	13
14	4.075	4.660	5.629	6.571	7.790	21.064	23.685	26.119	29.141	31.319	14
15	4.601	5.229	6.262	7.261	8.547	22.307	24.996	27.488	30.578	32.801	15
16	5.142	5.812	6.908	7.962	9.312	23.542	26.296	28.845	32.000	34.267	16
17	5.697	6.408	7.564	8.672	10.085	24.769	27.587	30.191	33.409	35.718	17
18	6.265	7.015	8.231	9.390	10.865	25.989	28.869	31.526	34.805	37.156	18
19	6.844	7.633	8.907	10.117	11.651	27.204	30.144	32.852	36.191	38.582	19
20	7.434	8.260	9.591	10.851	12.443	28.412	31.410	34.170	37.566	39.997	20
21	8.034	8.897	10.283	11.591	13.240	29.615	32.671	35.479	38.932	41.401	21
22	8.643	9.542	10.982	12.338	14.041	30.813	33.924	36.781	40.289	42.796	22
23	9.260	10.196	11.689	13.091	14.848	32.007	35.172	38.076	41.638	44.181	23
24	9.886	10.856	12.401	13.848	15.659	33.196	36.415	39.364	42.980	45.559	24
25	10.520	11.524	13.120	14.611	16.473	34.382	37.652	40.646	44.314	46.928	25

Continued

| | | | | | | α | | | | | |
ν	0.005	0.010	0.025	0.050	0.100	0.900	0.950	0.975	0.990	0.995	ν
26	48.290	45.642	41.923	38.885	35.563	17.292	15.379	13.844	12.198	11.160	26
27	49.645	46.963	43.195	40.113	36.741	18.114	16.151	14.573	12.879	11.808	27
28	50.993	48.278	44.461	41.337	37.916	18.939	16.928	15.308	13.565	12.461	28
29	52.336	49.588	45.722	42.557	39.087	19.768	17.708	16.047	14.256	13.121	29
30	53.672	50.892	46.979	43.773	40.256	20.599	18.493	16.791	14.953	13.787	30
35	60.275	57.342	53.203	49.802	46.059	24.797	22.465	20.569	18.509	17.192	35
40	66.766	63.691	59.342	55.758	51.805	29.051	26.509	24.433	22.164	20.707	40
45	73.166	69.957	65.410	61.656	57.505	33.350	30.612	28.366	25.901	24.311	45
50	79.490	76.154	71.420	67.505	63.167	37.689	34.764	32.357	29.707	27.991	50
55	85.749	82.292	77.380	73.311	68.796	42.060	38.958	36.398	33.570	31.735	55
60	91.952	88.379	83.298	79.082	74.397	46.459	43.188	40.482	37.485	35.534	60
65	98.105	94.422	89.177	84.821	79.973	50.883	47.450	44.603	41.444	39.383	65
70	104.215	100.425	95.023	90.531	85.527	55.329	51.739	48.758	45.442	43.275	70
75	110.286	106.393	100.839	96.217	91.061	59.795	56.054	52.942	49.475	47.206	75
80	116.321	112.329	106.629	101.879	96.578	64.278	60.391	57.153	53.540	51.172	80

Continued

ν	0.995	0.990	0.975	0.950	0.900	α 0.100	0.050	0.025	0.010	0.005	ν
85	55.170	57.634	61.389	64.749	68.777	102.079	107.522	112.393	118.236	122.325	85
90	59.196	61.754	65.647	69.126	73.291	107.565	113.145	118.136	124.116	128.299	90
95	63.250	65.898	69.925	73.520	77.818	113.038	118.752	123.858	129.973	134.247	95
100	67.328	70.065	74.222	77.929	82.358	118.498	124.342	129.551	135.807	140.169	100

Appendix 9

Equivalent Sigma Levels, Percent Defective, and PPM

	With no sigma shift (centered)				With 1.5 sigma shift		
Sigma level	Percent in specification	Percent defective	PPM	Sigma level	Percent in specification	Percent defective	PPM
0.10	7.9656	92.0344	920344	0.10	2.5957	97.40426	974043
0.20	15.8519	84.1481	841481	0.20	5.2235	94.77650	947765
0.30	23.5823	76.4177	764177	0.30	7.9139	92.08606	920861
0.40	31.0843	68.9157	689157	0.40	10.6950	89.30505	893050
0.50	38.2925	61.7075	617075	0.50	13.5905	86.40949	864095
0.60	45.1494	54.8506	548506	0.60	16.6196	83.38043	833804
0.70	51.6073	48.3927	483927	0.70	19.7952	80.20480	802048
0.80	57.6289	42.3711	423711	0.80	23.1240	76.87605	768760
0.90	63.1880	36.8120	368120	0.90	26.6056	73.39444	733944
1.00	68.2689	31.7311	317311	1.00	30.2328	69.76721	697672
1.10	72.8668	27.1332	271332	1.10	33.9917	66.00083	660083
1.20	76.9861	23.0139	230139	1.20	37.8622	62.13784	621378
1.30	80.6399	19.3601	193601	1.30	41.8185	58.18148	581815
1.40	83.8487	16.1513	161513	1.40	45.8306	54.16937	541694

Continued

Sigma level	With no sigma shift (centered)			Sigma level	With 1.5 sigma shift		
	Percent in specification	Percent defective	PPM		Percent in specification	Percent defective	PPM
1.50	86.6386	13.3614	133614	1.50	49.8650	50.13499	501350
1.60	89.0401	10.9599	109599	1.60	53.8860	46.11398	461140
1.70	91.0869	8.9131	89131	1.70	57.8573	42.14274	421427
1.80	92.8139	7.1861	71861	1.80	61.7428	38.25720	382572
1.90	94.2567	5.7433	57433	1.90	65.5085	34.49152	344915
2.00	95.4500	4.5500	45500	2.00	69.1230	30.87702	308770
2.10	96.4271	3.5729	35729	2.10	72.5588	27.44122	274412
2.20	97.2193	2.7807	27807	2.20	75.7929	24.20715	242071
2.30	97.8552	2.1448	21448	2.30	78.8072	21.19277	211928
2.40	98.3605	1.6395	16395	2.40	81.5892	18.41082	184108
2.50	98.7581	1.2419	12419	2.50	84.1313	15.86869	158687
2.60	99.0678	0.9322	9322	2.60	86.4313	13.56867	135687
2.70	99.3066	0.6934	6934	2.70	88.4917	11.50830	115083
2.80	99.4890	0.5110	5110	2.80	90.3191	9.68090	96809

Continued

	With no sigma shift (centered)				With 1.5 sigma shift		
Sigma level	Percent in specification	Percent defective	PPM	Sigma level	Percent in specification	Percent defective	PPM
2.90	99.6268	0.3732	3732	2.90	91.9238	8.07621	80762
3.00	99.7300	0.2700	2700	3.00	93.3189	6.68106	66811
3.10	99.8065	0.1935	1935	3.10	94.5199	5.48014	54801
3.20	99.8626	0.1374	1374	3.20	95.5433	4.45868	44567
3.30	99.9033	0.0967	967	3.30	96.4069	3.59311	35931
3.40	99.9326	0.0674	674	3.40	97.1283	2.87170	28717
3.50	99.9535	0.0465	465	3.50	97.7250	2.27504	22750
3.60	99.9682	0.0318	318	3.60	98.2135	1.78646	17865
3.70	99.9784	0.0216	216	3.70	98.6096	1.39035	13904
3.80	99.9855	0.0145	145	3.80	98.9276	1.07242	10724
3.90	99.9904	0.0096	96.2	3.90	99.1802	0.81976	8198
4.00	99.9937	0.0063	63.3	4.00	99.3790	0.62097	6210
4.10	99.9959	0.0041	41.3	4.10	99.5339	0.46612	4661
4.20	99.9973	0.0027	26.7	4.20	99.6533	0.34670	3467

Continued

With no sigma shift (centered)				With 1.5 sigma shift			
Sigma level	Percent in specification	Percent defective	PPM	Sigma level	Percent in specification	Percent defective	PPM
4.30	99.9983	0.0017	17.1	4.30	99.7445	0.25551	2555
4.40	99.9989	0.0011	10.8	4.40	99.8134	0.18858	1866
4.50	**99.9993**	**0.0007**	**6.8**	4.50	99.8650	0.13499	1350
4.60	99.9996	0.0004	4.2	4.60	99.9032	0.09676	968
4.70	99.9997	0.0003	2.6	4.70	99.9313	0.06871	687
4.80	99.9998	0.0002	1.6	4.80	99.9517	0.04834	483
4.90	99.99990	0.00010	1.0	4.90	99.9663	0.03369	337
5.00	99.99994	0.00006	0.6	5.00	99.9767	0.02326	233
5.10	99.99997	0.00003	0.3	5.10	99.9841	0.01591	159
5.20	99.99998	0.00002	0.2	5.20	99.9892	0.01078	108
5.30	99.999988	0.000012	0.12	5.30	99.9928	0.00723	72.3
5.40	99.999993	0.000007	0.07	5.40	99.9952	0.00481	48.1
5.50	99.999996	0.000004	0.04	5.50	99.9968	0.00317	31.7
5.60	99.999998	0.000002	0.02	5.60	99.9979	0.00207	20.7

Continued

	With no sigma shift (centered)				With 1.5 sigma shift		
Sigma level	Percent in specification	Percent defective	PPM	Sigma level	Percent in specification	Percent defective	PPM
5.70	99.9999988	0.0000012	0.012	5.70	99.9987	0.00133	13.3
5.80	99.9999993	0.0000007	0.007	5.80	99.9991	0.00085	8.5
5.90	99.9999996	0.0000004	0.004	5.90	99.9995	0.00054	5.4
6.00	99.9999998	0.0000002	0.002	6.00	99.9997	0.00034	3.4

Appendix 10

Glossary of Lean Six Sigma and Related Terms

ABC—*See* activity-based costing.

accuracy—The closeness of agreement between a test result or measurement result and the true or reference value.

activity-based costing (ABC)—A cost allocation method that allocates overhead expenses to activities based on the proportion of use, rather than proportion of costs.

appraisal costs—The costs associated with measuring, evaluating, or auditing products or services to ensure conformance to quality standards and performance requirements. These include costs such as incoming and source inspection/test of purchased material; in-process and final inspection/test; product, process, or service audit; calibration of measuring and test equipment; and the cost of associated supplies and materials.

ASQ—American Society for Quality.

assignable cause—A specifically identified factor that contributes to variation and is detectable. Eliminating assignable causes so that the points plotted on a control chart remain within the control limits helps achieve a state of statistical control. Note: Although *assignable cause* is sometimes considered synonymous with *special cause*, a special cause is assignable only when it is specifically identified. *See also* special cause.

attrition—Refers to loss of people to jobs outside the organization.

BB—*See* Black Belt.

benefit–cost analysis—A collection of the dollar value of benefits derived from an initiative divided by the associated costs incurred. A benefit–cost analysis is also known as a *cost–benefit analysis*.

bias—A systematic difference between the mean of the test result or measurement result and a true or reference value. For example, if one measures the lengths of 10 pieces of rope that range from 1 foot to 10 feet and always concludes that the length of each piece is 2 inches shorter than the true length, then the individual is exhibiting a bias of 2 inches. Bias is a component of accuracy.

Black Belt (BB)—A Lean Six Sigma role associated with an individual who is typically assigned full-time to train and mentor Green Belts as well as lead improvement projects using specified methodologies such as define, measure, analyze, improve, and control (DMAIC) and Design for Six Sigma (DFSS).

brainstorming—A problem-solving tool that teams use to generate as many ideas as possible that are related to a particular subject. Team members begin by offering all their ideas; the ideas are not discussed or reviewed until after the brainstorming session.

Certified Six Sigma Black Belt (CSSBB)—An ASQ certification.

Certified Six Sigma Green Belt (CSSGB)—An ASQ certification.

Certified Six Sigma Master Black Belt (CSSMBB)—An ASQ certification.

champion—A Lean Six Sigma role of a senior executive who ensures that his or her projects are aligned with the organization's strategic goals and priorities, provides the Lean Six Sigma team with resources, removes organizational barriers for the team, participates in project tollgate reviews, and essentially serves as the team's backer. Although many organizations define the terms "champion" and "sponsor" differently, they are frequently used interchangeably. *See also* sponsor.

chance cause—*See* random cause.

charter—A documented statement officially initiating the formation of a committee, team, project, or effort in which a clearly stated purpose and approval are conferred.

coaching—A process by which a more experienced individual helps enhance the existing skills and capabilities that reside in a less experienced individual. Coaching is about listening, observing, and providing constructive, practical, and meaningful feedback. Typically, coaching is used on a one-to-one basis, or for a small group or team, and conducted at the job site or during the training process. During training, coaching helps the trainee translate the theoretical learning into applied learning while helping the trainee develop confidence in their newly developing knowledge and skills. Post training, coaches help projects stay on track and advance toward completion in a timely manner.

common cause—*See* random cause.

communications plan—A document that defines what will be communicated, to whom, how often, and by what means.

control chart—A chart that plots a statistical measure of a series of samples in a particular order to steer the process regarding that measure and to control and reduce variation. The control chart comprises the plotted points, a set of upper and lower control limits, and a centerline. Specific rules are used to determine when the control chart goes out of control. Note: (1) the order is either time or sample number order based, and (2) the control chart operates most effectively when the measure is a process characteristic correlated with an ultimate product or service characteristic. The control chart is one of the seven basic tools of quality.

control limit—A line on a control chart used for judging the stability of a process. Note: (1) control limits provide statistically determined boundaries for the deviations from the centerline of the statistic plotted on a Shewhart control chart due to random causes alone; (2) control limits (with the exception of the acceptance control chart) are based on actual process data, not on specification limits; (3) other than points outside the control limits, "out-of-control" criteria can include runs, trends, cycles, periodicity, and unusual patterns within the control limits; (4) the calculation of control limits depends on the type of control chart.

control plan—A living document that identifies critical input or output variables and associated activities that must be performed to maintain control of the variation of processes, products, and services in order to minimize deviation from their preferred values.

COQ—*See* cost of quality.

cost avoidance (budget impacting)—This type of cost avoidance eliminates or reduces items in the budget marked for future spending. For example, three engineers are budgeted for hire in the fourth quarter. The cost of the three engineers is not in the baseline, nor has any spending for these engineers occurred. Eliminating these planned expenses has an impact on the future budget and is thus cost avoidance.

cost avoidance (non–budget impacting)—This type of cost avoidance results from productivity or efficiencies gained in a process (that is, reduction of non-value-added activities) without a head count reduction. For example, the process cycle time that involved two workers was reduced by 10 percent. Assuming that the 10 percent amounts to two hours per worker per week, the two workers save four hours per week. These four hours are allocated to other tasks.

cost avoidance savings—These are, by exclusion, everything that is not hard dollar savings. Cost avoidance savings are also known as *soft dollar savings*. There are two types of cost avoidance savings: budget impacting, and non–budget impacting.

cost–benefit analysis—*See* benefit–cost analysis.

cost of quality (COQ)—Costs specifically associated with the achievement or nonachievement of product or service quality, including all product or service requirements established by the organization and its contracts with customers and society. More specifically, quality costs are the total costs incurred by (1) investing in the prevention of nonconformances to requirements, (2) appraising a product or service for conformance to requirements, and (3) failing to meet requirements. These can then be categorized as prevention, appraisal, and failure costs. *See also* appraisal costs, failure costs, and prevention costs.

cost savings—For project savings to generate real cost savings, a prior baseline of spending must be established, the dollars must have been planned and in the budget, and the savings must affect the bottom line (that is, the profit and loss statement or balance sheet).

Cost savings are also known as *hard dollar savings*.

critical-to-quality (CTQ)—A characteristic of a product or service that is essential to ensure customer satisfaction.

CSSBB—*See* Certified Six Sigma Black Belt.

CSSGB—*See* Certified Six Sigma Green Belt.

CSSMBB—*See* Certified Six Sigma Master Black Belt.

CTQ—*See* critical-to-quality.

culture—Westcott (2006) refers to culture as "the collective beliefs, values, attitudes, manners, customs, behaviors, and artifacts unique to an organization."

customer, external—A person or organization that receives a product, service, or information but is not part of the organization supplying it. *See also* customer, internal.

customer, internal—The person or department that receives another person's or department's output (product, service, or information) within an organization. *See also* customer, external.

defect—The nonfulfillment of a requirement related to an intended or specified use for a product or service. In simpler terms, a defect is anything not done correctly the first time. Note: The distinction between the concepts *defect* and *nonconformity* is important because it has legal connotations, particularly those associated with product liability issues. Consequently, the term "defect" should be used with extreme caution. Also known as a *nonconformity*.

defective—A unit of product that contains one or more defects with respect to the quality characteristic(s) under consideration. Also known as a *nonconformance*.

defects per million opportunities (DPMO)— A measure of capability for discrete (attribute) data found by dividing the number of defects by the number of opportunities for defects, multiplied by a million. DPMO allows for comparison of different types of product. *See also* parts per million.

defects per unit (DPU)—A measure of capability for discrete (attribute) data found by dividing the number of defects by the number of units.

Design for Six Sigma (DFSS)—A structured methodology that focuses on designing new products or services with the intent of achieving Six Sigma quality levels. *See also* DMADV, DMADOV, DMEDI, and IDOV.

design of experiments (DOE, DOX)—The arrangement in which an experimental program is to be conducted, including the selection of factor combinations and their levels. Note: The purpose of designing an experiment is to provide the most efficient and economical methods of reaching valid and relevant conclusions from the experiment. The selection of the design is a function of many considerations, such as the type of questions to be answered, the applicability of the conclusions, the homogeneity of experimental units, the randomization scheme, and the cost to run the experiment. A properly designed experiment will permit meaningful interpretation of valid results.

DFSS—*See* Design for Six Sigma.

DMADOV—A structured DFSS methodology. DMADOV is an acronym for define, measure, analyze, design, optimize, and verify. Variations of DMADOV exist. *See also* Design for Six Sigma, DMADV, DMEDI, and IDOV.

DMADV—A structured DFSS methodology. DMADV is an acronym for define, measure, analyze, design, and verify. Variations of DMADV exist. *See* also Design for Six Sigma, DMADOV, DMEDI, and IDOV.

DMAIC—A structured methodology that focuses on improving existing processes with the intent of achieving Six Sigma quality levels. DMAIC is an acronym for define, measure, analyze, improve, and control.

DMEDI—A structured DFSS methodology. DMEDI stands for define, measure, explore, develop, and implement. Variations of DMEDI exist. *See also* Design for Six Sigma, DMADV, DMADOV, and IDOV.

DOE—*See* design of experiments.

DOX—*See* design of experiments.

DPMO—*See* defects per million opportunities.

DPU—*See* defects per unit.

driving forces—In the context of a force-field analysis, driving forces are those forces that aid in achieving the objective.

facilitator—An individual responsible for creating favorable conditions that will enable a team to reach its purpose or achieve its goals by bringing together the necessary tools, information, and resources to get the job done.

failure costs—The costs resulting from product or services not conforming to requirements or customer/user needs. Failure costs are further divided into internal and external categories. See also failure costs, external, and failure costs, internal.

failure costs, external—The failure costs occurring after delivery or shipment of the product, or during or after furnishing of a service, to the customer. Examples include the costs of processing customer complaints, customer returns, warranty claims, and product recalls.

failure costs, internal—The failure costs occurring prior to delivery or shipment of a product, or the furnishing of a service, to the customer. Examples include the costs of scrap, rework, reinspection, retesting, material review, and downgrading.

force-field analysis—A technique for analyzing the forces that aid or hinder an organization in reaching an objective.

gage repeatability and reproducibility (GR&R) study—A type of measurement system analysis done to evaluate the performance of a test method or measurement system. Such a study quantifies the capabilities and limitations of a measurement instrument, often estimating its repeatability and reproducibility. It typically involves multiple operators measuring a series of measurement items multiple times.

Gantt chart—A type of bar chart used in process/project planning and control to display planned work and finished work in relation to time. Also called a *milestone chart*.

gap analysis—A technique that compares an organization's existing state with its desired state (typically expressed by its long-term plans) to help determine what needs to be done to remove or minimize the gap.

GB—*See* Green Belt.

gemba visit—The term "gemba" means "place of work" or "the place where the truth can be found." Still others may call it "the value proposition." A gemba visit is a method of obtaining voice of the customer information that requires the design team to visit and observe how the customer uses the product in his or her environment.

governance—According to Bertin and Watson (2007), governance establishes the policy framework within which organization leaders will make strategic decisions to fulfill the organizational purpose, as well as the tactical actions that they take at the level of operational management to deploy and execute the organization's guiding policy and strategic direction. Governance in the context of Lean Six Sigma deployment includes all the processes, procedures, rules, roles, and responsibilities associated with the strategic, tactical, and operational deployment of Lean Six Sigma. It also includes the authoritative, decision-making body charged with the responsibility of providing the required governance.

GR&R—*See* gage repeatability and reproducibility (GR&R) study.

Green Belt (GB)—A Lean Six Sigma role associated with an individual who retains his or her regular position within the firm but is trained in the tools, methods, and skills necessary to conduct Lean Six Sigma improvement projects either individually or as part of larger teams.

hard dollar savings—*See* cost savings.

house of quality—A diagram named for its house-shaped appearance that clarifies the relationship between customer needs and product features. It helps correlate market or customer requirements and analysis of competitive products with higher-level technical and product characteristics, and makes it possible to bring several factors into a single figure. The house of quality is also known as *quality function deployment* (QFD).

IDOV—A structured methodology as an alternative to Design for Six Sigma. IDOV is an acronym for identify, design, optimize, and validate (sometimes shown as "verify"). Variations of IDOV exist. *See also* Design for Six Sigma, DMADV, DMADOV, and DMEDI.

in-control process—A condition where the existence of special causes is no longer indicated by a Shewhart control chart. This does not indicate that only random causes remain, nor does it imply that the distribution of the remaining values is normal (Gaussian). It indicates (within limits) a predictable and stable process.

intervention—An action taken by a leader or a facilitator to support the effective functioning of a team or work group.

Lean—A comprehensive approach complemented by a collection of tools and techniques that focus on reducing cycle time, standardizing work, and reducing waste. Lean is also known as *lean approach* or *lean thinking*.

lean approach—*See* Lean.

Lean Six Sigma—A fact-based, data-driven philosophy of improvement that values defect prevention over defect detection. It drives customer satisfaction and bottom-line results by reducing variation, waste, and cycle time while promoting the use of work standardization and flow, thereby creating a competitive advantage. It applies anywhere variation and waste exist, and every employee should be involved. Note: In the first edition of *The Certified Six Sigma Black Belt Handbook*, this definition was attributed to Six Sigma. However, through experience and empirical evidence, it has become clear that Lean and Six Sigma are opposite sides of the same coin. Both concepts are required to effectively drive sustained breakthrough improvement. Subsequently, the definition has been changed to reflect the symbiotic relationship that must exist between Lean and Six Sigma to ensure lasting and positive change.

lean thinking—*See* Lean.

Malcolm Baldrige National Quality Award (MBNQA)—An award established by Congress in 1987 to raise awareness of quality management and to recognize U.S. organizations that have implemented successful quality management systems. Criteria for Performance Excellence are published every other year. Up to three awards may be given annually in each of six categories: manufacturing businesses, service businesses, small businesses, education institutions, healthcare organizations, and nonprofit. The award is named after the late secretary of commerce Malcolm Baldrige, a proponent of quality management. The U.S. Commerce Department's National Institute of Standards and Technology manages the award, and ASQ administers it. The major emphasis in determining success is achieving results driven by effective processes.

Master Black Belt (MBB)—A Lean Six Sigma role associated with an individual typically assigned full-time to train and mentor Black Belts to ensure that improvement projects chartered are the right strategic projects for the organization. Master Black Belts are usually the authorizing body to certify Green Belts and Black Belts within an organization.

MBB—*See* Master Black Belt.

MBNQA—*See* Malcolm Baldrige National Quality Award.

measurement error—The difference between the actual value and the measured value of a quality characteristic.

mentoring—While coaching focuses on the individual as it relates to Lean Six Sigma, mentoring focuses on the individual from the career perspective. Mentors are usually experienced individuals (not necessarily in Lean Six Sigma) who have in-depth knowledge about the organization as well as the individual (that is, mentee). Usually, they come from within the organization, though not necessarily from the same department as their mentee. Their role is to help provide guidance, wisdom, and a possible road map to career advancement.

milestone chart—*See* Gantt chart.

natural process variation—*See* voice of the process.

noise factor—In robust parameter design, a noise factor is a predictor variable that is hard to control or is not desirable to control as part of the standard experimental conditions. In general, it is not desirable to make inference on noise factors, but they are included in an experiment to broaden the conclusions regarding control factors.

nonconformance—*See* defective.

nonconformity—*See* defect.

norms—Behavioral expectations, mutually agreed-on rules of conduct, protocols to be followed, or social practices.

objective—A quantitative statement of future expectations and an indication of when the expectations should be achieved; it flows from goal(s) and clarifies what people must accomplish.

observational study—Analysis of data collected from a process without imposing changes on the process.

operational definition—In the context of data collection and metric development, an operational definition is a clear, concise, and unambiguous statement that provides a unified understanding of the data for all involved before the data are collected or the metric developed. It answers "Who collects the data, how are the data collected, what data are collected, where is the source of the data, and when are the data collected?" Additional "who," "how," "what," "where," and "when" answers may be required. When the data are used to construct a metric, the operational definition further defines the formula that will be used and each term used in the formula. As with data collection, the operational definition also delineates who provides the metric and who is answerable to its results, how the metric is to be displayed or graphed, where the metric is displayed, and when it is available. Again, additional "who," "how," "what," "where," and "when" answers may be required. Finally, the operational definition provides an interpretation of the metric, such as "up is good."

out-of-control—A process is described as operating out of control when special causes are present.

parts per million (ppm)—The number of times an occurrence happens in one million chances. In a typical quality setting it usually indicates the number of times a defective part will happen in a million parts produced; the calculation is often projected into the future on the basis of past performance. Parts per million allows for comparison of different types of product. A ppm of 3.4 corresponds to a six sigma level of quality, assuming a 1.5 sigma shift of the mean.

ppm—*See* parts per million.

precision—The closeness of agreement between randomly selected individual measurements or test results. It is this aspect of measurement that addresses *repeatability*, or consistency when an identical item is measured several times.

prevention costs—The cost of all activities specifically designed to prevent poor quality in products or services. Examples include the cost of new product review, quality planning, supplier capability surveys, process capability evaluations, quality improvement team meetings, quality improvement projects, and quality education and training.

process capability—The calculated inherent variability of a characteristic of a product. It represents the best performance of the process over a period of stable operations. Process capability is expressed as $6\hat{\sigma}$, where $\hat{\sigma}$ is the sample standard deviation (short-term component of variation) of the process under a state of statistical control. Note: $\hat{\sigma}$ is often shown as σ in most process capability formulas. A process is said to be "capable" when the output of the process always conforms to the process specifications.

process mapping—The flowcharting of a process in detail that includes the identification of inputs and outputs as well as the categorization of inputs.

process owner—A Lean Six Sigma role associated with an individual who coordinates the various functions and work activities at all levels of a process, has the authority or ability to make changes in the process as required, and manages the entire process cycle so as to ensure performance effectiveness.

project—A project exists for producing results. Three components must be present for an activity to be defined as a project: scope, schedule, and resources.

project, active—Projects that have been assigned and are under way. Note: Projects that have been assigned, are under way, but have been placed on hold are considered active.

project, inactive—Projects that currently reside in the pipeline. These include both selected and prioritized projects.

project assignment—The process by which a project is assigned for execution and removed from the pipeline. A prioritized project and a resource (that is, Black Belt or Black Belt candidate) are available to be matched. If the matching is successful, the resource is assigned to the project and the project gets under way.

project closure—The process by which a project is closed and ready to be removed from active status. A project has been terminated or closed for one of two reasons: successful completion or it is no longer viable.

project identification—The process by which a project is identified as a possible improvement opportunity. The delineation of clear and specific improvement opportunities needed in the organization.

project management—The discipline of planning, organizing, and managing resources to bring about the successful completion of specific project goals and objectives, categorized into five component processes (sometimes called *life cycle phases*): initiating, planning, executing, controlling, and closing. The Project Management Institute notes that project management is the "application of knowledge, skills, tools, and techniques" to activities to meet the requirements of a specific project.

project prioritization—The process by which a project is prioritized once it has been selected and waiting in the project pipeline. A selected project will be independently assessed against specific criteria in order to establish a rank score. Rank scores of other selected projects will be compared, and a final ranking of the projects will be established. This final ranking will set the precedence and order for project assignment.

project qualification—The process by which a project is qualified for consideration as a Lean Six Sigma project. The project qualification process translates an identified improvement opportunity into Lean Six Sigma charter format for recognizing its completeness and the ability to assess its potential as a viable project.

project selection—The process by which a project is selected to enter the project pipeline. To select a project means that it has met the criteria of being a Lean Six Sigma project (that is, it has been qualified) and now must be judged to determine whether it is sufficiently worthy such that the organization is willing to invest time and resources to achieve the expected benefits of the project.

QFD—*See* quality function deployment.

quality council—A group within an organization that drives the quality improvement effort and usually has oversight responsibility for the implementation and maintenance of the quality management system; it operates in parallel with the normal operation of the business. A quality council is sometimes referred to as a *quality steering committee*.

quality function deployment (QFD)—A method used to translate voice of the customer information into product requirements/CTQs and to continue deployment (for example, cascading) of requirements to parts and process requirements. Quality function deployment is also known as the *house of quality*. *See also* house of quality.

random cause—A source of process variation that is inherent in a process over time. Note: In a process subject only to random cause variation, the variation is predictable within statistically established limits. Random cause is also known as *chance cause* and *common cause*.

rational subgroup—A subgroup (usually consecutive items) that is expected to be as free as possible from assignable causes. In a rational subgroup, variation is presumed to be only from random cause.

recognition—In the context of team growth, recognition is the sixth or final stage. *See also* stages of team growth.

restraining forces—In the context of a force-field analysis, restraining forces are those forces that hinder or oppose the objective.

revenue growth—In the context of Lean Six Sigma, the projected increase in income that will result from a project. This is calculated as the increase in gross income minus the cost. Revenue growth may be stated in dollars per year or as a percentage per year.

root cause—A factor (that is, original cause) that, through a chain of cause and effect, causes a defect or nonconformance to occur. Root causes should be permanently eliminated through process improvement. Note: Several root causes may be present and may work either together or independently to cause a defect.

sample—A group of units, portions of material, or observations taken from a larger collection of units, quantity of material, or observations that serves to provide information that may be used as a basis for making a decision concerning the larger quantity.

seven M's—The six M's with the addition of *management. See also* six M's.

seven types of waste—Taiichi Ohno proposed value as the opposite of waste and identified seven categories: (1) defects, (2) overproduction (ahead of demand), (3) overprocessing (beyond customer requirements), (4) waiting, (5) unnecessary motion, (6) transportation, and (7) inventory (in excess of the minimum).

simulation—The act of modeling the characteristics or behaviors of a physical or abstract system.

six M's—Typically, the primary categories of the cause-and-effect diagram: machines, manpower, materials, measurements, methods, and Mother Nature. Using these categories as a structured approach provides some assurance that few causes will be overlooked. *See also* seven M's.

Six Sigma—"Six Sigma" can take on various definitions across a broad spectrum depending on the level of focus and implementation. (1) Philosophy—the philosophical perspective views all work as processes that can be defined, measured, analyzed, improved, and controlled (DMAIC). Processes require inputs and produce outputs. If you control the inputs, you will control the outputs. This is generally expressed as the $Y = f(X)$ concept. (2) Set of tools—Six Sigma as a set of tools includes all the qualitative and quantitative techniques used by the Six Sigma practitioner to drive process improvement through defect reduction and the minimization of variation. A few such tools include statistical process control (SPC), control charts, failure mode and effects analysis (FMEA), and process mapping. There is probably little agreement among Six Sigma professionals as to what constitutes the tool set. (3) Methodology—this view of Six Sigma recognizes the underlying and rigorous approach known as DMAIC. DMAIC defines the steps a Six Sigma practitioner is expected to follow, starting with identifying the problem and ending with implementing long-lasting solutions. While DMAIC is not the only Six Sigma methodology in use, it is certainly the most widely adopted and recognized. (4) Metrics—in simple terms, Six Sigma quality performance means 3.4 defects per million opportunities (accounting for a 1.5-sigma shift in the mean).

SMART—SMART stands for specific, measurable, achievable, relevant, and timely, and is used generally in regard to the creation of goal statements.

soft dollar savings—*See* cost avoidance savings.

special cause—A source of process variation other than inherent process variation. Note: (1) sometimes *special cause* is considered synonymous with *assignable cause*, but a special cause is assignable only when it is specifically identified; (2) a special cause arises because of specific circumstances that are not always present. Therefore, in a process subject to special causes, the magnitude of the variation over time is unpredictable. *See also* assignable cause.

specification—An engineering requirement used for judging the acceptability of a particular product/service based on product characteristics such as appearance, performance, and size. In statistical analysis, specifications refer to the document that prescribes the requirements with which the product or service has to perform.

specification limits—Limiting value(s) stated for a characteristic. *See also* tolerance.

sponsor—A member of senior management who oversees, supports, and implements the efforts of a team or initiative. Although many organizations define the terms "champion" and "sponsor" differently, they are frequently used interchangeably. *See also* champion.

stable process—A process that is predictable within limits; a process that is subject only to random causes. (This is also known as a state of statistical control.) Note: (1) a stable process will generally behave as though the results are simple random samples from the same population; (2) this state does not imply that the random variation is large or small, within or outside specification limits, but rather that the variation is predictable using statistical techniques; (3) the process capability of a stable process is usually improved by fundamental changes that reduce or remove some of the random causes present and/or adjusting the mean toward the target value.

stages of team growth—The six development stages through which groups typically progress: forming, storming, norming, performing, adjourning, and recognition. Knowledge of the stages helps team members accept the normal problems that occur on the path from forming a group to becoming a team.

stakeholders—The people, departments, and groups that have an investment or interest in the success of or actions taken by the organization.

stratification—The layering of objects or data; also, the process of classifying data into subgroups based on characteristics or categories. Stratification is occasionally considered one of the seven tools of quality.

team—Two or more people who are equally accountable for the accomplishment of a purpose and specific performance goals; it is also defined as a small number of people with complementary skills who are committed to a common purpose.

team building—The process of transforming a group of people into a team and developing the team to achieve its purpose.

tolerance—The difference between upper and lower specification limits.

tollgate review—A formal review process conducted by a champion who asks a series of focused questions aimed at ensuring that the team has performed diligently during the current phase. The result of a tollgate is a "go" or "no-go" decision. The go decision allows the team to move forward to the next phase. If it is in the last phase, the go decision brings about project closure. If the decision is no-go, the team must remain in the phase or retreat to an earlier phase, or perhaps the project is terminated or suspended.

validation—This refers to the effectiveness of the design process itself and is intended to ensure that the design process is capable of meeting the requirements of the final product or process.

value stream mapping—A technique for following the production path of a product or service from beginning to end while drawing a visual representation of every process in the material and information flows. Subsequently, a future-state map is drawn of how value should flow.

verification—This refers to the design meeting customer requirements and ensures that the design yields the correct product or process.

virtual team—A boundaryless team functioning without a commonly shared physical structure or physical contact, using technology to link the team members. Team members are typically remotely situated and affiliated with a common organization, purpose, or project.

VOC—*See* voice of the customer.

voice of the customer (VOC)—Organizations apply significant effort to hear the voice of the customer by identifying the customers' needs and expectations. Once identified, organizations try to understand these needs and expectations and attempt to provide products and services that truly meet them.

voice of the process (VOP)—The 6σ spread between the upper and lower control limits as determined from an in-control process. The VOP is also known as *natural process variation*.

VOP—*See* voice of the process.

Y = f(X)—The foundational concept of Six Sigma. This equation represents the idea that process outputs are a function of process inputs.

Appendix 11

Glossary of Japanese Terms

baka-yoke—A term for a manufacturing technique for preventing mistakes by designing the manufacturing process, equipment, and tools so that an operation literally can not be performed incorrectly. In addition to preventing incorrect operation, the technique usually provides a warning signal of some sort for incorrect performance.

chaku-chaku—A term meaning "load-load" in a cell layout where a part is taken from one machine and loaded into the next.

gemba visit—A term that means "place of work" or "the place where the truth can be found." Still others may call it "the value proposition." A gemba visit is a method of obtaining voice of the customer information that requires the design team to visit and observe how the customer uses the product in his or her environment.

heijunka—The act of leveling the variety or volume of items produced at a process over a period of time. Heijunka is used to avoid excessive batching of product types and volume fluctuations, especially at a pacemaker process.

hoshin kanri—The selection of goals, projects to achieve the goals, designation of people and resources for project completion, and establishment of project metrics.

hoshin planning—A term meaning "breakthrough planning." Hoshin planning is a strategic planning process in which a company develops up to four vision statements that indicate where the company should be in the next five years. Company goals and work plans are developed based on the vision statements. Periodic audits are then conducted to monitor progress.

jidoka—A method of autonomous control involving the adding of intelligent features to machines to start or stop operations as control parameters are reached, and to signal operators when necessary. Jidoka is also known as *autonomation*.

kaikaku—A term meaning a breakthrough improvement in eliminating waste.

kaizen—A term that means gradual unending improvement by doing little things better and setting and achieving increasingly higher standards.

kaizen blitz/event—An intense team approach to employ the concepts and techniques of continuous improvement in a short time frame (for example, to reduce cycle time, increase throughput).

kanban—A system that signals the need to replenish stock or materials or to produce more of an item (also called a "pull" approach). The system was inspired by Taiichi Ohno's (Toyota) visit to a U.S. supermarket.

muda—An activity that consumes resources but creates no value; the seven categories are transportation, inventory, motion, waste, overproduction, overprocessing, and waiting.

poka-yoke—A term that means to mistake-proof a process by building safeguards into the system that avoid or immediately find errors. The term comes from the Japanese terms *poka*, which means "error," and *yokeru*, which means "to avoid."

seiban—The name of a management practice taken from the words *sei*, which means "manufacturing," and *ban*, which means "number."

A seiban number is assigned to all parts, materials, and purchase orders associated with a particular customer, job, or project. This enables a manufacturer to track everything related to a particular product, project, or customer and facilitates setting aside inventory for specific projects or priorities. Seiban is an effective practice for project and build-to-order manufacturing.

seiketsu—One of the five S's that means to conduct seiri, seiton, and seiso at frequent, indeed daily, intervals to maintain a workplace in perfect condition.

seiri—One of the five S's that means to separate needed tools, parts, and instructions from unneeded materials and to remove the latter.

seiso—One of the five S's that means to conduct a cleanup campaign.

seiton—One of the five S's that means to neatly arrange and identify parts and tools for ease of use.

shitsuke—One of the five S's that means to form the habit of always following the first four S's.

Bibliography

Andersen, Bjørn. 2007. *Business Process Improvement Toolbox*, 2nd ed. Milwaukee: ASQ Quality Press.

Anderson-Cook, Christine M., and Connie M. Borror. 2013. "Seven Data Collection Strategies to Enhance Your Quality Analyses." *Quality Progress*, April, 19–29.

Ashenbaum, Bryan. 2006. "Defining Cost Reduction and Cost Avoidance." Critical Issues Report. CAPS: Center for Strategic Supply Research.

AT&T Handbook Committee. 1958. *Statistical Quality Control Handbook*. Charlotte, NC: Western Electric.

Benbow, Donald W., and T. M. Kubiak. 2005. *The Certified Six Sigma Black Belt Handbook*. Milwaukee: ASQ Quality Press.

Bertin, Marcos E. J., and Gregory H. Watson, eds. 2007. *Corporate Governance: Quality at the Top*. Salem, NH: GOAL/QPC.

Covey, Stephen R. 1989. *The 7 Habits of Highly Effective People*. New York: Simon and Schuster.

Kubiak, T. M. 2004. "What Does It Take to Be a Master Black Belt?" *Six Sigma Forum Magazine*, February, 40–41.

Kubiak, T. M. 2007. "Reviving the Process Map." *Quality Progress*, May, 59–63.

Kubiak, T. M. 2008. "Data Dependability." *Quality Progress*, June, 61–64.

Kubiak, T. M., and Donald W. Benbow. 2009. *The Certified Six Sigma Black Belt Handbook*, 2nd ed. Milwaukee: ASQ Quality Press.

Kubiak, T. M. 2012. *The Certified Six Sigma Master Black Belt Handbook*. Milwaukee: ASQ Quality Press.

Kubiak, T. M. 2012. "Operational Definitions: The Lifeblood of Data Collection and Metric Development." *Quality Progress*, August, 51–53.

Lynch, Donald P., Suzanne Bertolino, and Elaine Cloutier. 2003. "How to Scope DMAIC Projects." *Quality Progress*, January, 37–41.

Minitab 15 software. 2007. Minitab Inc., State College, PA.

Montgomery, Douglas C. 2005. *Introduction to Statistical Quality Control*, 5th ed. Hoboken, NJ: John Wiley & Sons.

Munro, Roderick A., Matthew J. Maio, Mohamed B. Nawaz, Govindarajan Ramu, and Daniel J. Zrymiak. 2008. *The Certified Six Sigma Green Belt Handbook*. Milwaukee: ASQ Quality Press.

Phillips, Jack J., and Ron Drew Stone. 2002. *How to Measure Training Effectiveness: A Practical Guide to Tracking the Six Key Indicators*. New York: McGraw-Hill.

Portny, Stanley E. 2010. *Project Management for Dummies*. Hoboken, NJ: Wiley.

Price, Mark, and James Works. 2010. "Balancing Roles and Responsibilities in Six Sigma." www.isixsigma.com. Accessed 10/4/13. http://www.isixsigma.com/tools-templates/raci-diagram/balancing-roles-and-responsibilities-six-sigma/.

Rooney, James J., T. M. Kubiak, Russ Westcott, R. Dan Reid, Keith Wagoner, Peter E. Pylipow, and Paul Plsek. 2009. "Building from the Basics." *Quality Progress*, January, 21–22.

Siebels, Donald L. 2004. *The Quality Improvement Glossary*. Milwaukee: ASQ Quality Press.

Tague, Nancy R. 2005. *The Quality Toolbox*, 2nd ed. Milwaukee: ASQ Quality Press.

Westcott, Russell T., ed. 2014. *The Certified Manager of Quality/Organizational*

Excellence Handbook, 3rd ed. Milwaukee: ASQ Quality Press.

Wheeler, Donald J., and David S. Chambers. 1992. *Understanding Statistical Process Control*, 2nd ed. Knoxville, TN: SPC Press.

Woodford, David. 2010. "Design for Six Sigma—IDOV Methodology." isixsigma.com. Accessed 10/4/13. http://www.isixsigma.com/new-to-six-sigma/design-for-six-sigma-dfss/design-six-sigma-idov-methodology/.

Index

NOTES

NOTES

NOTES

NOTES

www.ingramcontent.com/pod-product-compliance
Lightning Source LLC
Chambersburg PA
CBHW061237220326
41599CB00028B/5447